Contents

national
STATISTICS

Department for
Transport

Focus on Ports

2006 edition

palgrave
macmillan

First published 2006 by
PALGRAVE MACMILLAN
Houndmills, Basingstoke, Hampshire RG21 6XS and
175 Fifth Avenue, New York, N.Y. 10010
Companies and representatives throughout the world

PALGRAVE MACMILLAN is the global academic imprint of the Palgrave Macmillan
division of St. Martin's Press, LLC and of Palgrave Macmillan Ltd. Macmillan® is a
registered trademark in the United States, United Kingdom and other countries.
Palgrave is a registered trademark in the European Union and other countries.

ISBN-13: 978–0–230–00215–9
ISBN-10: 0–230–00215–3

This book is printed on paper containing at least 75% recycled fibre.

A catalogue record for this book is available from the British Library.

10 9   8   7   6   5   4   3   2   1
15 14 13 12 11 10 09 08 07 06

Printed and bound in Great Britain by William Clowes Ltd, Beccles, Suffolk

Prepared by the Maritime Statistics Branch, Department for Transport
Alan Brown, John Gathercole, Nazir Pathan, Steve Wellington

Enquiries about maritime statistics
Department for Transport
2/19 Great Minster House
76 Marsham Street
London SW1P 4DR
Telephone 020 7944 3087
Web site www.dft.gov.uk/transtat
Email: maritime.stats@dft.gsi.gov.uk

DfT is often prepared to sell unpublished data. Further information can be obtained
from the above address.

List of tables

List of figures

Symbols and conventions used

Areas covered: except where otherwise stated, statistics refer to the United Kingdom.

Rounding of figures: figures in the tables have been independently rounded and therefore totals may differ from the constituent items.

Symbols: .. not available
– nil or negligible (less than half the final digit shown).

Foreword

Focus on Ports 2006 is an updated and revised version of the first edition published in November 2000. This edition provides a compilation of data from the latest *Maritime Statistics* and other sources, together with an authoritative commentary. It features a breakdown of market sectors and types of cargo handled by ports, gives facts and figures about individual ports around the coast, and includes a chapter on port employment and accident rates. It is a valuable reference document for those interested in the UK ports industry.

For further information about the report or contents please contact: Steve Wellington, Department for Transport, 2/19, Great Minster House, 76 Marsham Street, London SW1P 4 DR Tel: 020 7944 4131 email: maritime.stats@dft.gov.uk.

The following periodical statistics are also produced by the Department:

1. Maritime Statistics
Annual compilation of statistics about freight and passenger traffic through UK ports. The report includes statistics about UK-owned and registered ships and the world fleet.

2. Fleet Circulars
Quarterly statistical summaries of UK registered and UK-owned shipping, and half-yearly statistical summaries of the world fleet.

3. Waterborne Freight in the United Kingdom
Annual report of freight traffic moved within the UK by water transport, covering coastwise and inland waters traffic.

4. Sea Passenger Bulletins
Statistical summaries of international and domestic sea passengers to and from UK ports, by main routes.

If you would also like to receive copies of any of these circulars or reports, please contact the Department at the above address.

Introduction to the UK ports industry

Background

This report updates the November 2000 edition of *Focus on Ports*, which was published in parallel with *Modern Ports: A UK Policy*. Information in this latest report is based principally on *Maritime Statistics 2004* and earlier annual volumes, together with supporting information from other available sources.

The importance of shipping and trade to the economy of the UK, an island nation, has resulted in the establishment of a large number of ports around the coast, which are very diverse in terms of size and type of cargo handled. In total there are more than 650 ports in the UK for which statutory harbour authority powers have been granted, of which around 120 are commercially active. They range from ports such as the Port of London, which extends 95 miles from Teddington to the North Sea, to small harbour trusts responsible for quays, piers and other facilities which are only of local significance.

It is estimated that around 95 per cent by volume and 75 per cent by value of the UK's international trade is transported by sea. In 2004, total UK imports across all transport modes were valued at £249 billion and exports at £191 billion, which indicates that approximately £330 billion of the UK's international trade was moved through its seaports.

The UK ports industry is the largest in Europe in terms of freight tonnage, handling a total of 573 million tonnes of foreign and domestic traffic in 2004. Figure 1.1 shows the distribution of port traffic at UK ports. In addition, each year around 50 million international and domestic passenger journeys are made through UK ports. In 2004, there were 27 million international ferry and cruise passenger journeys to and from the UK, a further 4 million domestic passengers on sea crossings and 19 million on inter-island services such as the Isle of Wight and Scottish islands.

Annual tonnage handled by UK ports grew steadily between 1980 and 2000 at around 1.3 per cent per year. Traffic levels declined during the period 2000 to 2003 by 1 per cent annually but rose again by 3 per cent in 2004. Total, inward and outward traffic movements are shown in Figure 1.2, whilst Figure 1.3 presents the same information but in terms of imports, exports and domestic traffic. Growth in imports has been much stronger than exports over the last 20 years, reflecting the changing structure of the economy from manufacturing to service industries. Domestic traffic has declined over the same period.

Growth in traffic has been particularly strong in two key sectors, container and roll-on/roll-off (ro-ro) traffic, which have averaged growths of 5 per cent and 3.5 per cent per year respectively over the last decade. Container and ro-ro traffic trends are shown in Figure 1.4.

FIGURE 1.1 Traffic through UK ports, 2004

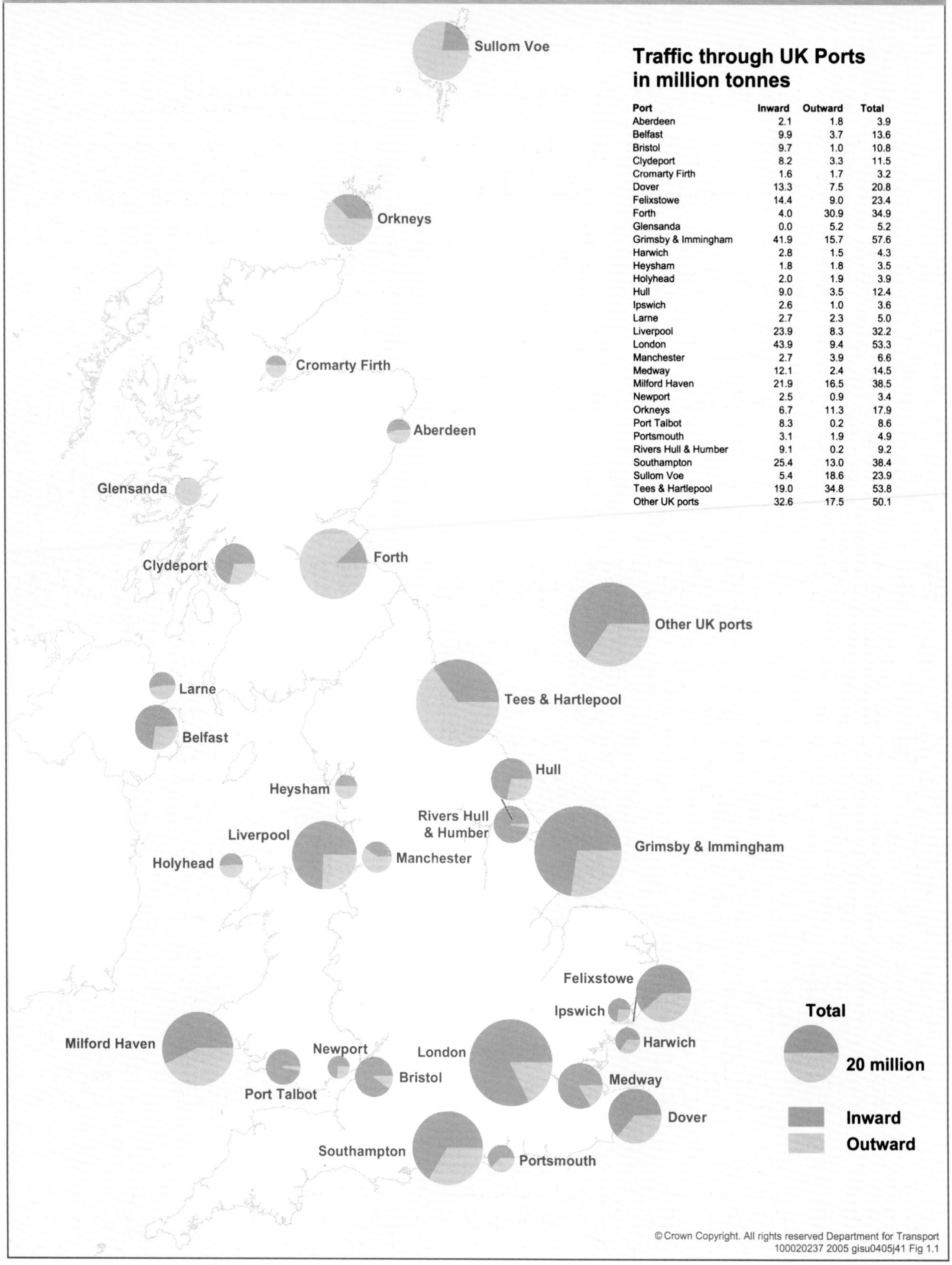

Traffic through UK Ports in million tonnes

Port	Inward	Outward	Total
Aberdeen	2.1	1.8	3.9
Belfast	9.9	3.7	13.6
Bristol	9.7	1.0	10.8
Clydeport	8.2	3.3	11.5
Cromarty Firth	1.6	1.7	3.2
Dover	13.3	7.5	20.8
Felixstowe	14.4	9.0	23.4
Forth	4.0	30.9	34.9
Glensanda	0.0	5.2	5.2
Grimsby & Immingham	41.9	15.7	57.6
Harwich	2.8	1.5	4.3
Heysham	1.8	1.8	3.5
Holyhead	2.0	1.9	3.9
Hull	9.0	3.5	12.4
Ipswich	2.6	1.0	3.6
Larne	2.7	2.3	5.0
Liverpool	23.9	8.3	32.2
London	43.9	9.4	53.3
Manchester	2.7	3.9	6.6
Medway	12.1	2.4	14.5
Milford Haven	21.9	16.5	38.5
Newport	2.5	0.9	3.4
Orkneys	6.7	11.3	17.9
Port Talbot	8.3	0.2	8.6
Portsmouth	3.1	1.9	4.9
Rivers Hull & Humber	9.1	0.2	9.2
Southampton	25.4	13.0	38.4
Sullom Voe	5.4	18.6	23.9
Tees & Hartlepool	19.0	34.8	53.8
Other UK ports	32.6	17.5	50.1

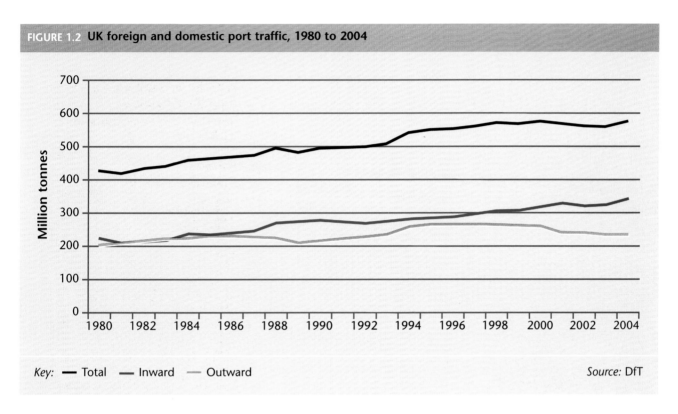

FIGURE 1.2 UK foreign and domestic port traffic, 1980 to 2004

Key: — Total — Inward — Outward

Source: DfT

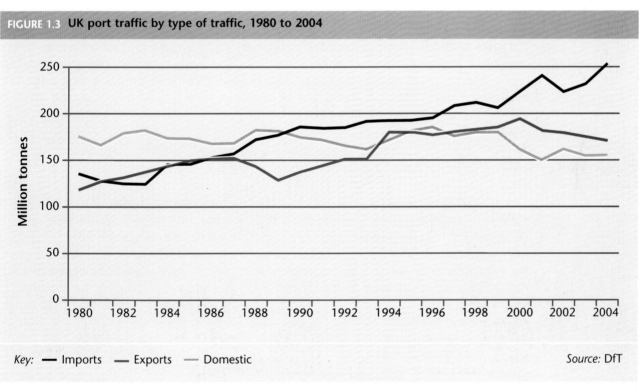

FIGURE 1.3 UK port traffic by type of traffic, 1980 to 2004

Key: — Imports — Exports — Domestic

Source: DfT

Much of the UK's total tonnage is concentrated in a relatively small number of ports – the top 15 ports account for almost 80 per cent of the UK's total port traffic. Table 1.1 shows a ranking of ports handling 2.0 million tonnes or more of cargo annually in 2004. Table 1.2 shows freight traffic at the top UK ports compared with the largest ports in Northern Europe. Grimsby & Immingham, the largest port in the UK, is the sixth largest port in Northern Europe, whilst Tees & Hartlepool and London follow in seventh and eighth places respectively.

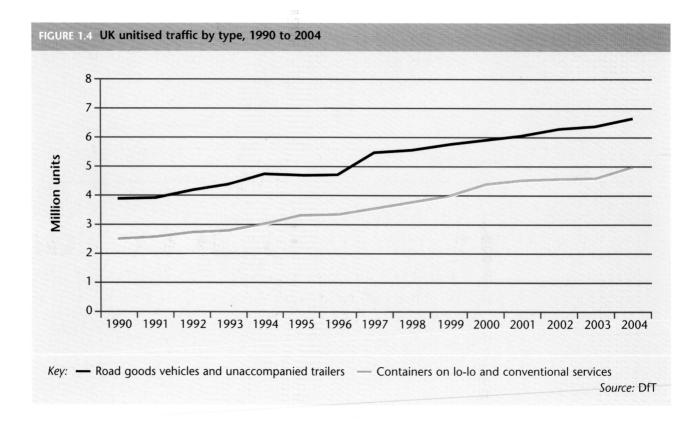

FIGURE 1.4 UK unitised traffic by type, 1990 to 2004

Key: ▬ Road goods vehicles and unaccompanied trailers ▬ Containers on lo-lo and conventional services

Source: DfT

UK port traffic is also very concentrated regionally – almost a third of tonnage goes through south-east ports, as shown in Figure 1.5. The regional concentration of container traffic is even more pronounced – three-quarters of container units were handled at south-east ports in 2004. The two busiest container ports outside the south-east in 2004 were Liverpool and Hull (Figure 1.6).

Port management

Port management is carried out by companies, trusts and (or on behalf of) municipal authorities and other public bodies. With the exception of naval dockyards, they operate independently of government. The various business activities going on within ports such as cargo handling, however, may be carried out by several different bodies. At some ports the port authority is responsible for all the port's business activities within its jurisdiction. At other ports the authority may only be responsible for some functions such as conservancy and pilotage (this is the case for the Port of London Authority, Harwich Haven Authority and Falmouth Harbour commissioners). In Scotland there are a large number of 'marine works' which provide facilities for inter-island ferry services. Some ports may have fishery interests with some operating only as fishery harbours. Appendices B to E provide maps of UK ports and harbours according to their status (that is, private, trust or publicly managed).

Company and private ports

The private sector operates all but 5 of the 20 largest ports by tonnage and handles around two-thirds of the UK's port traffic. Some company ports, such as Felixstowe and Manchester, have always been in private hands. A group of ports formerly operated by the

TABLE 1.1 Ports handling at least 2 million tonnes of cargo, 2004

Rank	Port	Million tonnes	Rank	Port	Million tonnes
1	Grimsby & Immingham	57.6	19	Manchester	6.6
2	Tees & Hartlepool	53.8	20	Glensanda	5.2
3	London	53.3	21	Larne	5.0
4	Milford Haven	38.5	22	Portsmouth	4.9
5	Southampton	38.4	23	Harwich	4.3
6	Forth	34.9	24	Holyhead	3.9
7	Liverpool	32.2	25	Aberdeen	3.9
8	Sullom Voe	23.9	26	Ipswich	3.6
9	Felixstowe	23.4	27	Heysham	3.5
10	Dover	20.8	28	Newport, Gwent	3.4
11	Orkney	17.9	29	Cromarty Firth	3.2
12	Medway	14.5	30	Tyne	3.0
13	Belfast	13.6	31	Cairnryan	2.8
14	Hull	12.4	32	Cardiff	2.5
15	Clydeport	11.5	33	Trent River	2.3
16	Bristol	10.8	34	Goole	2.2
17	River Hull & Humber	9.2	35	Plymouth	2.2
18	Port Talbot	8.6	36	Warrenpoint	2.0

Source: DfT

TABLE 1.2 Northern Europe's largest cargo ports, 1990 to 2004

	Million tonnes									
	1990	1992	1994	1996	1998	2000	2001	2002	2003	2004
Rotterdam	288	292	293	284	307	320	314	321	327	352
Antwerp	102	104	110	107	120	131	130	132	143	152
Hamburg	61	65	68	71	76	86	93	98	107	115
Le Havre	54	53	54	56	66	67	69	68	71	76
Amsterdam	47	49	48	55	56	64	68	70	65	74
Grimsby & Immingham	39	41	43	47	48	53	55	56	56	58
Tees & Hartlepool	40	43	43	45	51	52	51	50	54	54
London	58	49	52	53	57	48	51	51	51	53
Bremen	30	31	31	32	35	45	46	47	49	52
Dunkirk	37	40	37	35	39	45	45	48	50	51

Source: DfT, ISL and Port of Rotterdam

British Transport Docks Board were privatised in the 1980s, becoming Associated British Ports (ABP). ABP currently own 21 ports which account for 25 per cent of UK traffic. Several major former trust ports – Clydeport, Dundee, Forth, Ipswich, Sheerness (part of Medway), Tees, Hartlepool and Tilbury (part of London) were privatised between 1992 and 1997 under the Ports Act 1991. Appendix C provides a map of UK ports in the company-owned sector.

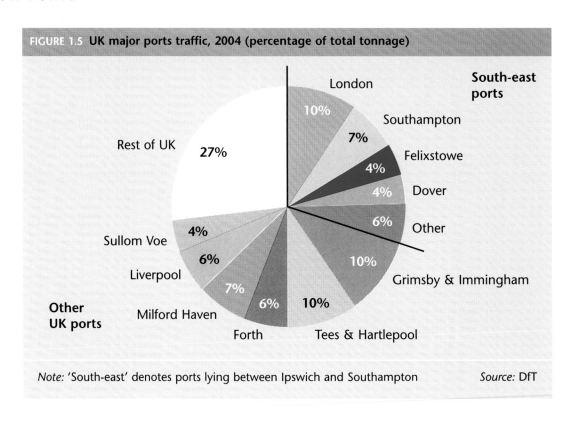

FIGURE 1.5 UK major ports traffic, 2004 (percentage of total tonnage)

Note: 'South-east' denotes ports lying between Ipswich and Southampton   *Source:* DfT

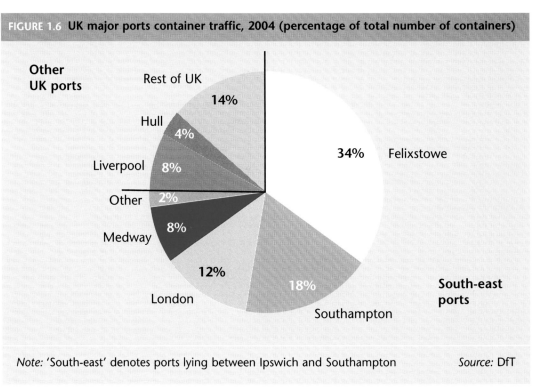

FIGURE 1.6 UK major ports container traffic, 2004 (percentage of total number of containers)

Note: 'South-east' denotes ports lying between Ipswich and Southampton   *Source:* DfT

Trust ports

There are around 100 trust ports in the UK, half of which are in England and Wales. They are mainly small to medium-sized enterprises (around a quarter have an annual turnover exceeding £1 million). The most significant trust ports are the Port of London Authority, Aberdeen, Belfast, Dover and Milford Haven.

Trust ports are independent statutory bodies governed by independent boards of trustees acting in the interests of all stakeholders. Any surplus revenue from operations is normally ploughed back into improving facilities. Some trust port authorities, such as the Port of London Authority and Harwich Haven Authority, act as conservancy and pilotage authorities but do not run or operate terminal facilities. Some of the biggest trust ports support the fishing industry, whilst others focus on leisure activities. Some trust ports have income derived solely from non-port-related activities. Ports in general are now more commonly leasing land for business use and residential developments. Appendix D provides a map of trust ports in the UK. In 2004 there were 14 trust ports with turnover above £3 million, of which 8 were above £10 million.

Municipal and other publicly owned ports

Many municipal ports concentrate on leisure and fishing activities. There are, however, several that have significant volumes of port traffic. These include Sullom Voe (Shetland), Flotta (Orkney), Portsmouth, Ramsgate and Sunderland. Sullom Voe is one of the newest harbour authorities, established to serve the North Sea oil terminal in the Shetland Isles. Appendix E provides a map of municipal and other publicly owned ports in the UK.

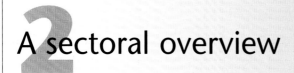 A sectoral overview

Introduction

The UK ports industry is very diverse, catering for different markets and reflecting different geographical locations and local conditions. This section discusses the different market sectors and types of cargo handled by ports. The main sectors, types of cargo and other activities at UK ports comprise:

- liquid bulk traffic
- dry bulk traffic
- container traffic
- roll-on/roll-off freight and passengers
- general cargo
- cruise passengers
- fishing
- inland waters
- recreation.

Traffic through UK ports can also be divided into foreign traffic (traffic to and from ports outside the UK), coastwise traffic (traffic between UK ports) and 'one-port' traffic (traffic to and from UK offshore oil and gas installations plus sea-dredged aggregates). Domestic traffic is the sum of coastwise and one-port traffic.

Historical comparisons

Traffic trends over the past 30 years are summarised first by broad cargo types and whether the traffic is foreign, coastwise or one port. Subsequent sections give more detail about each cargo type. Interpretation of these longer term series, however, needs to take account of changes in data collection and coverage over the period. The main changes over the period are as follows:

- Before 1980, statistics for Northern Ireland were unavailable and totals before 1980 are for Great Britain, not the UK. Breakdown by individual cargo type for Northern Ireland was not available until 1988.
- From 1995, traffic broken down by cargo type is only available for major ports. Since, however, major ports account for about 95 per cent of all UK port traffic, the change has not resulted in significant discontinuities.
- More substantial changes to data collection were made in 2000 to meet the requirements of the EC Maritime Statistics Directive. Changes were made to the definitions of individual cargo categories, and major ports were defined as those handling 1 million tonnes or more of traffic (whereas previously it was 2 million tonnes). More

detail was collected about particular types of traffic (for example containers) and new information was collected about the port of origin and port of destination. Overall, the change did not result in significant discontinuities, although for some traffic types, notably containers, the 1999 and 2000 results are not entirely comparable.

A breakdown of UK port tonnage since 1985 by main cargo category is shown in Table 2.1 and Figure 2.1. Liquid bulk is the most dominant cargo type, accounting for just under half of all traffic in 2004, although its relative share of all traffic has dropped from around 60 per cent in 1985. Dry bulk traffic has increased by around a third and now accounts for about one-fifth of traffic. The most significant changes have occurred in container and ro-ro traffic; container traffic has more than doubled over the period and ro-ro traffic is now more than two and a half times the level in 1985. In market share terms, container and ro-ro traffic have both seen their share of total traffic increase (from 5 to 10 per cent and 7 to 16 per cent respectively) over the period.

TABLE 2.1 UK port traffic by cargo category, 1985 to 2004								
	Million tonnes							
	1985	**1990**	**1995**	**2000**	**2001**	**2002**	**2003**	**2004**
Liquid bulk	273	256	290	294	277	273	263	269
Dry bulk	93	118	116	114	124	116	124	123
Containers[1]	24	35	48	52	52	51	51	57
Ro-ro	33	52	66	86	84	88	88	93
Other cargo	26	31	29	28	30	30	29	31
Total	**449**	**492**	**548**	**573**	**566**	**558**	**556**	**573**

1 Table excludes some container traffic travelling on ro-ro vessels after 1995

Source: DfT

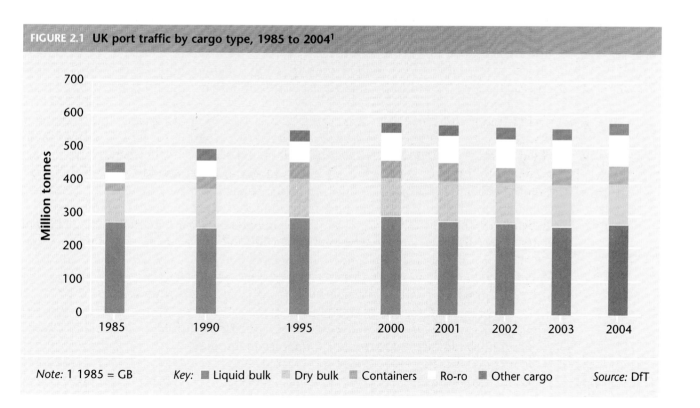

FIGURE 2.1 UK port traffic by cargo type, 1985 to 2004[1]

Note: 1 1985 = GB   *Key:* ■ Liquid bulk   Dry bulk   Containers   ■ Ro-ro   ■ Other cargo   *Source:* DfT

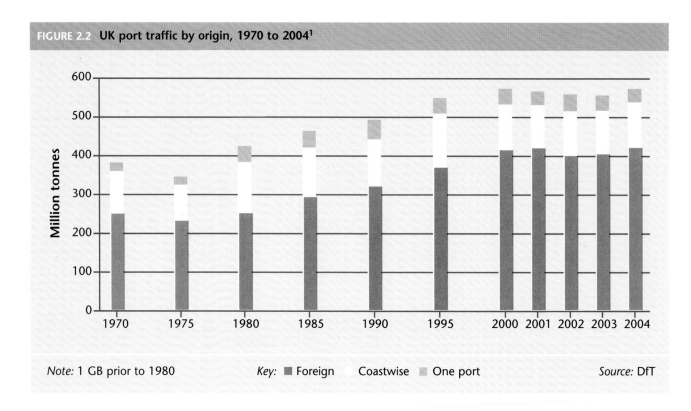

FIGURE 2.2 UK port traffic by origin, 1970 to 2004[1]

Note: 1 GB prior to 1980   *Key:* ■ Foreign   □ Coastwise   ▨ One port   *Source:* DfT

Foreign, coastwise and one-port traffic shares since 1970 are given in Figure 2.2. Foreign traffic, the largest component, has risen from 244 million tonnes in 1970 to 420 million tonnes in 2004, around two-thirds of which was with Europe and the Mediterranean region. Proportionally, the share of foreign traffic has risen from 66 per cent to 73 per cent of all UK traffic, reflecting the growing importance of international trade. Coastwise traffic rose during the late 1970s and again during the early 1990s but has declined to around the 1970 level over recent years (118 million tonnes in 2004). Given that overall traffic levels have been increasing, the coastwise share has fallen from 29 per cent in 1970 to 21 per cent in 2004. One-port traffic (mainly from offshore oil and gas terminals) increased to a peak of 50 million tonnes in 1990 before falling back. However, the current level of 35 million tonnes is almost double that of 1970, with market share currently at 6 per cent.

Liquid bulk traffic

Liquid bulk traffic peaked in 2000, since when it has fallen by around 8 per cent – see Table 2.1 and Figure 2.1. Liquid bulk is, in tonnage terms, the largest cargo category at UK ports, accounting for 47 per cent of all traffic (269 million tonnes) in 2004. The two largest components at major ports were crude oil (162 million tonnes) and oil products (86 million tonnes) and key port facilities have been developed to handle the large tankers in which oil is transported. Chemicals account for a large proportion of miscellaneous liquid bulk.

Liquid bulk cargoes account for just over 40 per cent of imports and exports, and a much higher proportion (around two-thirds) of domestic shipping.

Crude oil

Much of the growth in crude oil traffic occurred during the late 1970s when domestic traffic increased with the start of oil production in the North Sea, and the decline in traffic since 2000 reflects the recent fall in UK production. Around three-quarters of UK North Sea oil is transported from the production fields by pipeline, with the remainder delivered by tanker (one-port traffic) (Table 2.2 and Figure 2.3).

TABLE 2.2 UK crude oil traffic, 1965 to 2004[1]				
	Million tonnes			
Year	Foreign	Coastwise	One port	Total
1965	64.7	1.5	–	66.2
1970	102.7	1.0	–	103.7
1975	90.2	2.7	0.5	93.4
1980	100.3	49.2	11.4	160.9
1985	105.0	48.5	13.2	166.7
1990	93.0	37.1	13.9	144.0
1995	115.7	56.7	10.8	183.3
2000	122.9	36.8	24.6	184.3
2001	114.6	36.5	17.5	168.6
2002	107.0	40.5	25.0	172.5
2003	99.7	38.4	22.2	160.3
2004	101.0	42.0	18.6	161.6
1 GB prior to 1990, major ports from 1995				
Source: DfT				

In 2004, 43 per cent of the UK's total crude oil traffic was handled at three ports, Tees & Hartlepool (25 million tonnes), Sullom Voe (24 million) and Forth (21 million). Some ports such as Sullom Voe and Flotta (Orkney) are almost entirely devoted to this activity (Table 2.3).

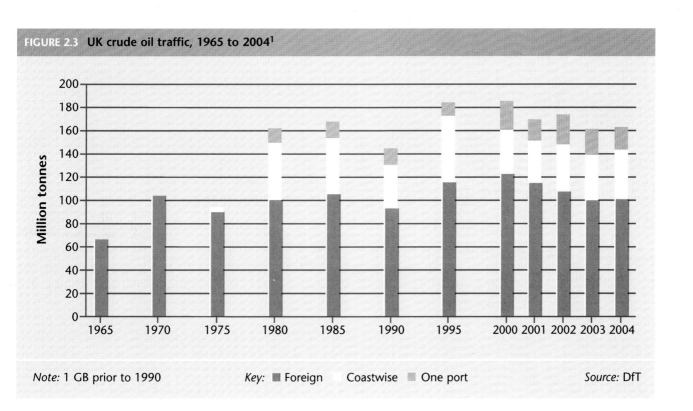

FIGURE 2.3 UK crude oil traffic, 1965 to 2004[1]

Note: 1 GB prior to 1990   *Key:* ■ Foreign   □ Coastwise   ▨ One port   *Source:* DfT

TABLE 2.3 Crude oil traffic by major UK port, 2004				
	Million tonnes			
	Foreign in	**Foreign out**	**Domestic**	**Total**
Tees & Hartlepool	0.3	13.2	11.8	25.3
Sullom Voe	0.2	16.9	6.7	23.9
Forth	–	15.5	5.4	20.9
Orkney	3.0	9.6	5.1	17.7
Southampton	8.4	0.5	7.7	16.7
Milford Haven	8.8	–	6.8	15.6
Liverpool	7.7	–	3.7	11.4
Grimsby & Immingham	3.2	–	5.9	9.2
River Hull & Humber	6.8	–	1.7	8.5
London	3.9	–	3.7	7.7
Cromarty Firth	0.4	1.4	1.1	2.9
Clydeport	0.8	–	0.7	1.5
Dundee	0.3	–	–	0.3
Total	**43.9**	**57.1**	**60.6**	**161.6**

Source: DfT

Oil products and liquid gas

The oil products and liquid gas category, which includes aviation and motor spirits, kerosene and liquid gas, has fallen from 107 million tonnes at its peak in 1970 to 93 million tonnes in 2004. Whilst coastwise traffic has halved since 1965, foreign traffic has doubled and is now around two-thirds of total traffic (Figure 2.4).

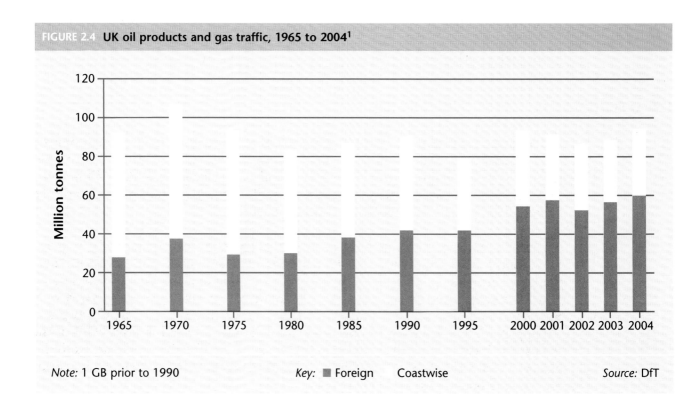

FIGURE 2.4 **UK oil products and gas traffic, 1965 to 2004[1]**

Note: 1 GB prior to 1990   *Key:* ■ Foreign   Coastwise   *Source:* DfT

TABLE 2.4 Oil products and liquified gas traffic by major UK port, 2004				
	Million tonnes			
	Foreign in	Foreign out	Domestic	Total
Milford Haven	4.0	10.6	7.2	21.9
Grimsby & Immingham	3.5	6.8	3.7	14.0
London	7.6	2.1	2.3	12.0
Southampton	3.4	5.5	1.4	10.3
Forth	0.7	4.7	4.4	9.7
Tees & Hartlepool	2.0	2.5	2.1	6.6
Manchester	0.6	1.9	1.0	3.6
Belfast	0.6	–	2.4	2.9
Bristol	0.9	–	1.5	2.4
Clydeport	0.5	0.3	1.2	2.0
Aberdeen	–	–	1.4	1.4
Plymouth	0.1	–	1.1	1.2
Medway	1.0	–	0.2	1.2
Cardiff	–	–	1.1	1.2
Hull	0.3	0.1	0.3	0.7
Sunderland	–	–	0.4	0.4
Ipswich	–	–	0.3	0.3
Peterhead	–	–	0.2	0.2
Londonderry	–	–	0.2	0.2
Shoreham	0.1	–	0.1	0.2
Other ports	0.1	0.2	0.7	1.0
Total	**25.4**	**34.7**	**33.3**	**93.4**

Source: DfT

A number of ports handle oil products, but the majority of traffic (73 per cent) in 2004 was handled at Milford Haven, Grimsby & Immingham, London, Southampton and Forth (Table 2.4).

Dry bulk traffic

Dry bulk cargo represented just over a fifth of all traffic in 2004 (123 million tonnes). The components of dry bulk at major ports were agricultural products (13 million tonnes), coal (41 million), ores (18 million) and miscellaneous dry bulk (42 million) (Figure 2.5). In 2004, 40 per cent of all dry bulk traffic was handled at Grimsby & Immingham, London and Tees & Hartlepool (Table 2.5).

Some bulk cargo ports have declined in recent years with the loss of their main business, for example ports exporting coal. Other ports have diversified and developed new markets. Ports which are situated close to either the origin or destination of bulk cargoes have an advantage over other ports but this advantage is increasingly being eroded with the move to larger ships. Such ships can only be received by larger ports which have also developed the bulk-handling facilities able to receive them. Inland infrastructure is also important since ships are likely to use ports which have railheads or good access to the primary road network. Some traditional ports handling bulk cargoes are now too small for their customers' ships.

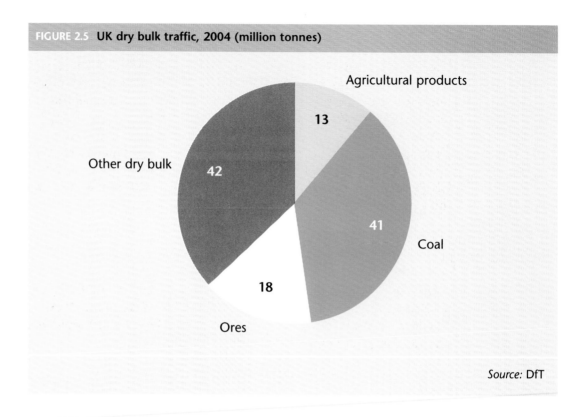

FIGURE 2.5 UK dry bulk traffic, 2004 (million tonnes)

Source: DfT

TABLE 2.5 Significant dry bulk cargoes by major UK port, 2004

Million tonnes							
Agricultural products		**Coal**		**Ores**		**Miscellaneous dry bulk**	
London	2.5	Grimsby & Immingham	11.0	Grimsby & Immingham	6.1	London	9.7
Liverpool	2.3	Clydeport	6.0	Tees & Hartlepool	5.7	Glensanda	5.2
Belfast	1.8	Tees & Hartlepool	4.7	Port Talbot	5.2	Liverpool	3.4
Grimsby & Immingham	0.9	Bristol	3.7	Tyne	0.5	Medway	2.9
Bristol	0.8	Medway	3.4	Trent River	0.2	Tees & Hartlepool	2.1
Southampton	0.7	Liverpool	3.0	Aberdeen	0.1	Belfast	1.7
Ipswich	0.7	Port Talbot	2.9	Hull	0.1	Bristol	1.5
Hull	0.4	London	2.0			Fowey	1.3
River Hull & Humber	0.4	Hull	1.5			Southampton	1.3
Clydeport	0.3	Londonderry	0.5			Grimsby & Immingham	1.2
Other ports	2.2	Other ports	2.5	Other ports	–	Other ports	11.7
Total	**12.9**	**Total**	**41.3**	**Total**	**17.8**	**Total**	**42.1**

Source: DfT

Ores

Traffic in ores totalled 18 million tonnes in 2004, most of which was imported from abroad. The main ports are those situated near traditional iron and steel centres. In 2004, three ports were responsible for 95 per cent of major port traffic – Grimsby & Immingham (6.1 million tonnes), Tees & Hartlepool (5.7 million) and Port Talbot (5.2 million). Prior to 2000, this category also included scrap metal which is now classified as miscellaneous dry bulk (Table 2.5).

Coal

Coal traffic remains important – 41 million tonnes in 2004 compared to 44 million tonnes in 1965. However, the UK has changed from being a net exporter of coal to a net importer, with outward traffic accounting for 60 per cent of all traffic in 1965 compared to 6 per cent in 2004 (Figure 2.6).

Grimsby & Immingham (11 million tonnes), Clydeport (6.0 million) and Tees & Hartlepool (4.7 million) all had major volumes of coal traffic in 2004 (Table 2.5). Only Clydeport had any significant outward traffic, with around 1.0 million tonnes being shipped to power stations in Northern Ireland.

Agricultural products

Traffic in agricultural products, such as cereals, fruit, animal feeds, sugars, oils and nuts, totalled 13 million tonnes in 2004. This was 13 per cent lower than in 1985 (Table 2.7). The leading ports for this traffic in 2004 were London (2.5 million tonnes), Liverpool (2.3 million) and Belfast (1.8 million) (Table 2.5).

Miscellaneous dry bulk

The miscellaneous dry bulk category, which includes sea-dredged aggregates, crude minerals, scrap metal, cement, chemicals and fertilisers, is the largest component of dry bulk. In 2004, this category totalled 42 million tonnes (Table 2.5).

Sea-dredged aggregates received at UK ports totalled 13 million tonnes in 2004. Traffic rose steadily during the late 1980s to reach a peak of 20 million tonnes but has since

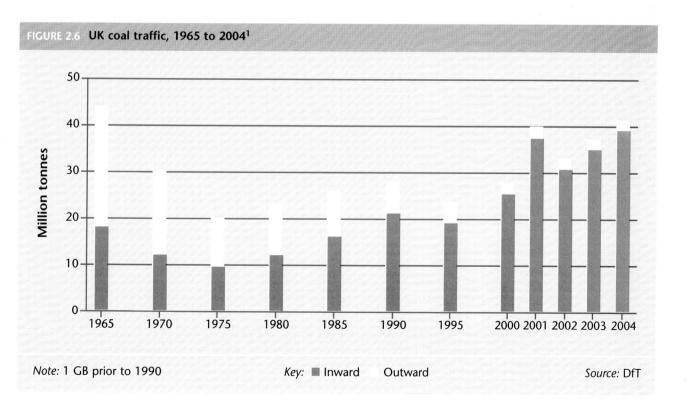

FIGURE 2.6 UK coal traffic, 1965 to 2004[1]

Note: 1 GB prior to 1990   Key: ■ Inward   Outward   Source: DfT

TABLE 2.6 UK sea-dredged aggregates, 1970 to 2004[1]	
	Million tonnes
Year	**Inward**
1970	9.1
1975	12.4
1980	13.0
1985	14.1
1990	19.5
1995	13.0
2000	11.1
2001	12.8
2002	13.8
2003	12.9
2004	12.6

1 GB prior to 1990. Major ports from 1995

Source: DfT

TABLE 2.7 UK agricultural products port traffic, 1985 to 2004[1]	
	Million tonnes
1985	14.8
1990	16.9
1995	15.6
2000	14.1
2001	13.0
2002	13.1
2003	14.8
2004	12.9

1 Major ports from 1995; GB prior to 1990

Source: DfT

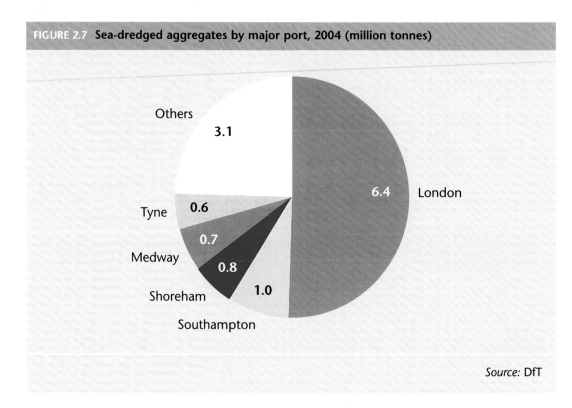

FIGURE 2.7 Sea-dredged aggregates by major port, 2004 (million tonnes)

Source: DfT

fallen back to the present level (Table 2.6). London is the largest port handling this traffic, accounting for half of major port traffic of this type in 2004 (6.4 million tonnes) (Figure 2.7).

Container traffic

Traffic in containers can be measured in tonnage terms in order to compare it with other types of cargo. However, it is more usually measured in terms of the number of units (20 ft, 40 ft and so on), or in terms of twenty foot equivalent units (TEU) where a 40 ft container is the equivalent of two units.

The majority of containers (about 90 per cent in 2004) are recorded as 'lift-on/lift-off' (lo-lo), where the container is lifted on or off a conventional container ship by crane or gantry. The remainder are carried on roll-on/roll-off (ro-ro) vessels and are excluded from the tables and charts from 2000 onward unless stated otherwise. Containers transported on lorries (including those pulling trailers and semi-trailers), however, are not counted as containers at all in the statistics, but as ro-ro vehicle traffic.

It is also important to note how container movements are treated; a container is counted each time it is transferred from the ship to the quay or from the quay to the ship. So, for example, a container coming from China into Southampton is counted once when it is unloaded at Southampton. If the container is then moved (transhipped) to Cardiff by ship, the same container is counted again when it is loaded onto a ship at Southampton, and then again when it is unloaded at Cardiff. The container is, therefore, counted three times in the statistics – both into and out of Southampton and once into Cardiff. No separate figures are available for the number of container transhipments.

In 2004, around 10 per cent of UK port tonnage was transported in containers. Over the past 15 years, there has been a substantial growth in both tonnage and number of containers handled by UK ports. In terms of TEU, container traffic through UK ports has increased by 112 per cent since 1990, rising from 4.0 million TEU to 8.4 million TEU in 2004, including those transported on ro-ro vessels (Figure 2.8).

Between 1990 and 2000, year-on-year growth in UK container traffic averaged around 6 per cent but has since slowed to an average of 4.5 per cent per annum (Figure 2.9). Most container traffic (around 80 per cent in 2004) is concentrated at five leading UK ports. Current shares are: Felixstowe (34 per cent), Southampton (18 per cent), London (12 per cent), Medway (8 per cent) and Liverpool (8 per cent) (Table 2.9a).

The largest growth of the leading ports since 2000 has been at London (averaging 15 per cent per year) and Southampton (8 per cent per year). Traffic through Felixstowe has

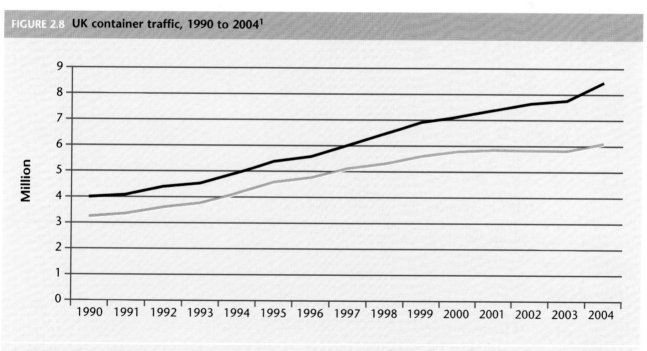

FIGURE 2.8 UK container traffic, 1990 to 2004[1]

Note: 1 Includes containers on ro-ro vessels which have been estimated from 2000 onward

Key: —— TEU —— Loaded TEU

Source: DfT

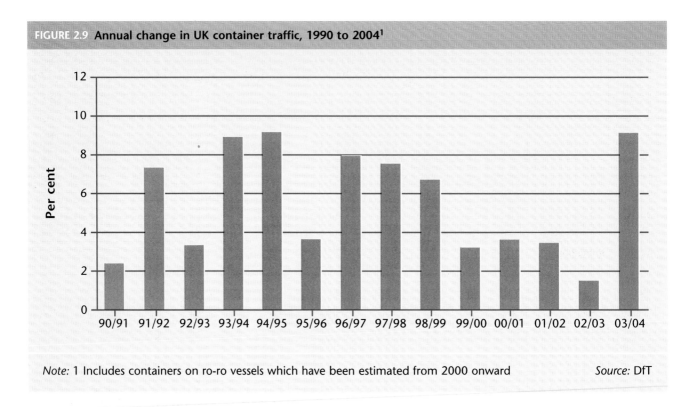

FIGURE 2.9 Annual change in UK container traffic, 1990 to 2004[1]

Note: 1 Includes containers on ro-ro vessels which have been estimated from 2000 onward   *Source:* DfT

TABLE 2.8 UK container traffic, 1985 to 2004[1,2]

	Tonnage (millions)	Loaded containers (millions)	Empty containers (millions)	All containers (millions)	TEUs (millions)	Tonnage per loaded container	% of empty containers
1985	23.7	1.68	0.45	2.13	3.05	14.1	21
1990	34.5	2.31	0.53	2.84	3.97	14.9	19
1995	47.6	3.10	0.54	3.64	5.36	15.4	15
2000	51.6	3.48	0.84	4.32	6.71	14.8	19
2001	51.7	3.51	0.94	4.45	6.98	14.7	21
2002	51.1	3.42	1.07	4.49	7.22	14.9	24
2003	51.3	3.25	1.26	4.51	7.30	15.8	28
2004	56.4	3.52	1.38	4.90	7.99	16.0	28

1 Excludes some container traffic travelling on ro-ro vessels from 2000; 2 Major ports from 2000

Source: DfT

fallen on average by 1 per cent over the same period (Table 2.9a). A relatively large number of container movements through UK ports are empty – 28 per cent in 2004 compared to 19 per cent in 2000. These were largely outward and reflect the trade imbalance between the UK and its main trading partners (Table 2.8).

In 2004, Felixstowe was the fifth largest container port in Northern Europe, handling 2.7 million TEU. The leading container ports were Rotterdam (8.3 million), Hamburg (7.0 million), Antwerp (6.1 million) and Bremen (3.5 million) (Table 2.9b).

The UK's conventional container traffic can be split into three distinct geographical sectors: short sea, deep sea and coastal traffic. Short sea traffic (trade with Europe and the Mediterranean) accounted for 39 per cent of total container traffic in 2004 (1.9 million

TABLE 2.9a **Container volumes at the leading UK container ports, 2000 to 2004[1]**

	2000 TEU (000s)	% of total	2001 TEU (000s)	% of total	2002 TEU (000s)	% of total	2003 TEU (000s)	% of total	2004 TEU (000s)	% of total
Felixstowe	2,825	42	2,839	41	2,684	37	2,482	34	2,717	34
Southampton	1,079	16	1,170	17	1,275	18	1,374	19	1,446	18
London	568	8	752	11	873	12	911	12	979	12
Medway	515	8	493	7	530	7	518	7	632	8
Liverpool	523	8	512	7	487	7	566	8	603	8
Hull	257	4	204	3	157	2	267	4	310	4
Belfast	193	3	176	3	187	3	210	3	228	3
Forth	137	2	154	2	176	2	187	3	215	3
Grimsby & Immingham	65	1	69	1	180	2	125	2	137	2
Tees & Hartlepool	42	1	79	1	125	2	133	2	133	2
Other major ports	505	8	533	8	547	8	529	7	592	7
Total major ports	**6,710**	**100**	**6,981**	**100**	**7,221**	**100**	**7,302**	**100**	**7,993**	**100**

1 Excludes some container traffic travelling on ro-ro vessels

Source: DfT

TABLE 2.9b **Container volumes at the major NW European container ports, 2000 to 2004**

	Thousand TEU				
	2000	2001	2002	2003	2004
Rotterdam	6,274	6,096	6,515	7,100	8,281
Hamburg	4,248	4,689	5,374	6,138	7,003
Antwerp	4,082	4,218	4,777	5,445	6,063
Bremen	2,737	2,915	2,999	3,191	3,469
Felixstowe	2,825	2,839	2,684	2,482	2,717
Le Havre	1,464	1,523	1,720	1,977	2,132

Source: DfT, ISL and Port of Rotterdam

units). Deep sea traffic (trade with all other foreign countries) accounted for 54 per cent of total traffic (2.6 million units), compared to 46 per cent in 1990. In 2004, about half of deep sea container traffic was with China, Hong Kong, Taiwan, Singapore and Japan – of which three-quarters of loaded container units were inward. Coastwise traffic (UK coastal movements) accounted for just 0.23 million units (5 per cent) in 2004. In 3 per cent of cases, the route was unknown (Table 2.10 and Figure 2.10).

Hub and feeder ports

The latest generation of container ships has a capacity of almost 10,000 TEU. As ships have become larger, shipping lines have reduced the number of ports they serve directly, leading to the use of hub ports. This process has generated a substantial business in the transhipment of containerised freight and is a key element of the throughput at some UK ports. A transhipment port such as Felixstowe may operate feeder services to other UK ports (for example Forth or Tyne) that are not on a major shipping route. Conversely, container traffic from Rotterdam and Antwerp, for example, may be transhipped on to UK ports.

TABLE 2.10 Container traffic by route, 1990 to 2004[1]

Area	1990	1993	1995	2000	2001	2002	2003	2004
				Thousand units				
All short sea	**1,291**	**1,293**	**1,522**	**1,764**	**1,763**	**1,799**	**1,802**	**1,900**
Belgium	181	200	216	196	210	290	352	349
France	20	22	49	91	91	75	91	105
Germany	104	78	65	182	186	208	184	206
Ireland	66	55	110	99	93	94	84	78
Netherlands	441	470	472	331	408	455	473	492
Scandinavia and Baltic (inc. Denmark)	221	234	244	287	254	210	204	232
Spain and Portugal	119	107	115	200	196	170	171	152
Other short sea	138	127	250	378	325	297	243	285
All deep sea	**1,315**	**1,586**	**1,865**	**2,227**	**2,284**	**2,319**	**2,368**	**2,636**
North America	407	418	458	559	543	471	421	432
Far East	492	636	763	1,042	1,071	1,202	1,324	1,513
Africa	117	175	218	166	172	150	147	154
Indian Ocean and the Gulf	124	135	236	227	241	247	264	309
Australasia	86	121	70	78	101	91	41	82
Other deep sea	88	101	121	155	156	158	171	144
Unspecified	–	–	–	109	122	115	100	129
All foreign (inc. unspecified)	**2,606**	**2,879**	**3,387**	**4,100**	**4,170**	**4,233**	**4,270**	**4,665**
Coastwise	**234**	**234**	**200**	**221**	**276**	**260**	**240**	**234**
Total	**2,840**	**3,113**	**3,587**	**4,321**	**4,446**	**4,493**	**4,510**	**4,899**

1 Excludes some container traffic travelling on ro-ro vessels from 2000

Source: DfT

FIGURE 2.10 UK container traffic by sea area, 1990 to 2004[1]

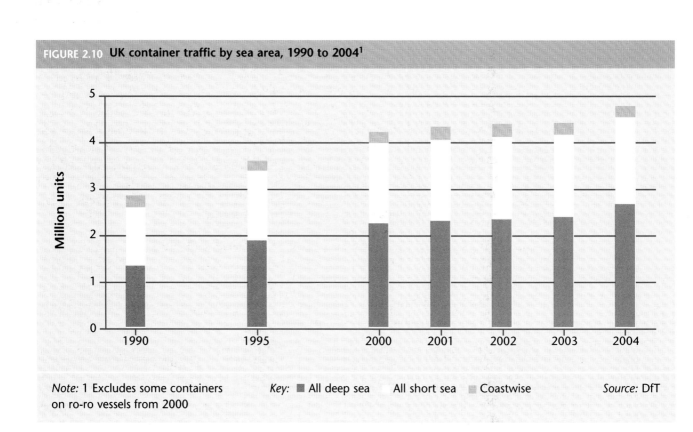

Note: 1 Excludes some containers on ro-ro vessels from 2000

Key: ■ All deep sea   □ All short sea   ▨ Coastwise

Source: DfT

TABLE 2.11 Container ship arrivals at UK ports, 2004			
		No. of vessels	
Deadweight tonnes	<20,000	20,000+	All
Felixstowe	813	1,634	2,447
Southampton	26	642	668
Medway	68	487	555
London	724	414	1,138
Liverpool	321	133	454
Grimsby & Immingham	332	–	332
Belfast	216	10	226
Forth	482	–	482
Hull	422	–	422
Tees & Hartlepool	404	1	405
Other ports	875	75	950
Total	**4,683**	**3,396**	**8,079**

Source: Lloyds Marine Intelligence Unit

With modern container ships carrying up to 10,000 TEU, vessel size is another factor that has to be taken into account when deciding on which port to unload a cargo. In 2004, Felixstowe, Southampton, Medway and London received almost 95 per cent of calls by container ships of 20 thousand deadweight tonnes or more (Table 2.11).

Roll-on/roll-off traffic

Freight

Ro-ro cargoes transported by road goods vehicles and unaccompanied trailers accounted for 14 per cent (81 million tonnes) of total UK port tonnage in 2004. The total of 6.5 million units represented an increase of 170 per cent since 1985. The ro-ro market in the UK is divided into a number of regional sectors. The leading ports in 2004 were Dover, London and Portsmouth (English Channel), Grimsby & Immingham, Harwich and Felixstowe (North Sea) and Liverpool, Larne and Belfast (Irish Sea) (Table 2.12 and Figure 2.11). Dover is by far the largest port for ro-ro freight traffic, with 2.0 million units in 2004 (30 per cent of the sector excluding the Channel Tunnel). The next busiest were Liverpool (0.43 million units) and Grimsby & Immingham (0.42 million) (Table 2.12).

The opening of the Channel Tunnel in 1994 had a significant impact on ro-ro ports. With traffic levels in 2004 of 1.3 million road goods vehicles, the Channel Tunnel had a larger share of the market than any port except for Dover. In 1995, the first full year after it opened, Channel Tunnel traffic made up 8 per cent of the total road goods vehicle market, rising to 16 per cent in 2004 (Figure 2.11).

Fortunes have varied amongst those ports competing for ro-ro business. Since 1990 (the period for which there are also figures for NI), the largest increases have been at Dover (1.0 million), Liverpool (0.34 million), Grimsby & Immingham (0.26 million) and Belfast (0.25 million). The strongest percentage growth since 2000 has been at Harwich (105 per cent), Ramsgate (71 per cent) and Holyhead (47 per cent) (Table 2.12).

TABLE 2.12 Road goods vehicles and unaccompanied trailers,[1] 1985 to 2004

				Units (000s)				
	1985	1990	1995	2000	2001	2002	2003	2004
Dover	719	988	1,043	1,618	1,771	1,854	1,785	1,981
Liverpool	69	94	221	403	438	415	392	433
Grimsby & Immingham	122	167	203	307	385	390	453	422
Larne	–	333	375	301	317	345	339	365
Harwich	139	176	163	177	182	238	303	363
Belfast	–	79	167	349	332	306	321	331
London	6	188	251	345	281	368	333	326
Portsmouth	103	173	302	292	290	289	297	291
Holyhead	34	44	98	185	208	215	231	272
Heysham	36	96	177	259	257	253	324	240
Felixstowe	177	187	332	445	372	329	232	226
Cairnryan	81	94	126	157	165	179	193	211
Hull	61	105	151	161	146	145	147	152
Ramsgate	77	164	261	83	95	135	147	143
Stranraer	115	150	143	155	139	122	117	127
Fleetwood	131	104	90	116	125	120	126	125
Other ports	539	696	544	439	659	407	467	507
Total by sea[2]	**2,409**	**3,837**	**4,648**	**5,792**	**6,161**	**6,110**	**6,206**	**6,515**
Channel Tunnel freight vehicles	–	–	**391**	**1,133**	**1,198**	**1,231**	**1,285**	**1,281**
Total incl. Channel Tunnel	**2,409**	**3,837**	**5,039**	**6,925**	**7,359**	**7,341**	**7,491**	**7,796**

1 Trailers accompanying powered road goods vehicles are not counted separately. Excludes shipborne port-to-port trailers and barges
2 Major ports only from 2000

Source: DfT

FIGURE 2.11 Road goods vehicles through UK ports, 1985 to 2004[1]

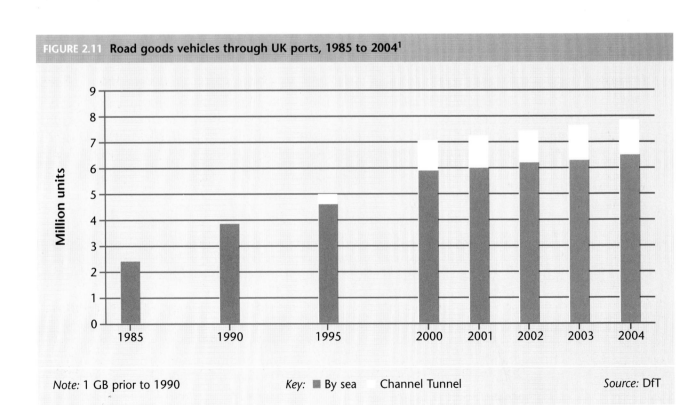

Note: 1 GB prior to 1990   *Key:* ■ By sea   Channel Tunnel   *Source:* DfT

Ferry passengers

Passenger services also serve different regional markets. In the English Channel, there are Straits of Dover services to France and Belgium, and western Channel services to France, Spain and the Channel Islands. There are also west coast services across the Irish Sea and

TABLE 2.13	UK international sea passengers by port, 1975 to 2004									
					Sea passengers (000s)					
	1975	1980	1985	1990	1995	2000	2001	2002	2003	2004
Short sea passengers										
Eastern Channel	..	..	17,677	19,900	22,487	16,922	16,301	16,838	15,176	14,798
Medway	..	453	677	800	78	–	–	–	–	–
Ramsgate	..	1,183	1,021	1,459	2,807	76	88	117	137	148
Dover	6,783	10,965	13,754	15,573	17,850	16,078	15,857	16,329	14,631	14,275
Folkestone	1,592	1,625	1,361	1,235	725	440	5	–	–	–
Newhaven	600	797	862	832	979	313	337	379	397	361
Other ports	..	..	2	1	48	15	14	13	11	14
Western Channel	..	..	2,332	3,992	4,817	4,274	4,513	4,665	4,357	4,238
Portsmouth	..	722	1,728	2,890	3,331	3,176	3,344	3,406	3,116	3,077
Southampton	1,140	1,073	–	–	531	–	–	–	–	5
Poole	..	42	59	455	373	455	586	620	623	520
Weymouth	..	206	169	8	–	60	–	8	15	20
Plymouth	..	322	376	639	582	583	583	631	603	617
West coast	..	..	2,790	2,729	3,598	4,234	3,882	3,880	3,802	3,656
Swansea	..	–	1	72	163	124	122	121	118	116
Milford Haven	..	289	239	247	341	463	388	387	384	378
Fishguard	280	430	529	757	945	832	687	662	645	614
Holyhead	791	1,142	1,594	1,622	2,125	2,518	2,380	2,371	2,333	2,262
Liverpool	385	561	421	31	24	293	298	291	269	270
Other ports	..	..	5	–	–	4	8	48	53	15
East coast	..	..	3,232	3,418	3,420	3,086	3,056	3,342	3,188	3,106
Lerwick	..	..	5	7	3	6	6	7	13	14
Forth	..	–	–	–	–	–	–	105	195	192
Tyne	129	179	74	203	406	667	745	816	829	767
Hull	292	484	443	981	961	972	1,006	1,041	994	976
Grimsby & Immingham	..	..	4	9	9	12	13	38	43	43
Great Yarmouth	104	9	11	3	–	–	–	–	–	
Ipswich	..	3	2	6	1	5	6	6	6	7
Felixstowe	299	998	764	445	447	86	80	58	19	19
Harwich	1,489	1,669	1,926	1,763	1,582	1,335	1,196	1,268	1,085	1,085
Other ports	..	..	3	2	11	3	4	3	4	3
All short sea passengers	15,556	23,395	26,029	30,044	34,322	28,516	27,753	28,726	26,523	25,798
Cruise and other long sea passengers	338	226	179	175	240	487	496	572	726	807
All international passengers	15,894	23,621	26,208	30,219	34,562	29,003	28,249	29,298	27,249	26,605

Source: DfT

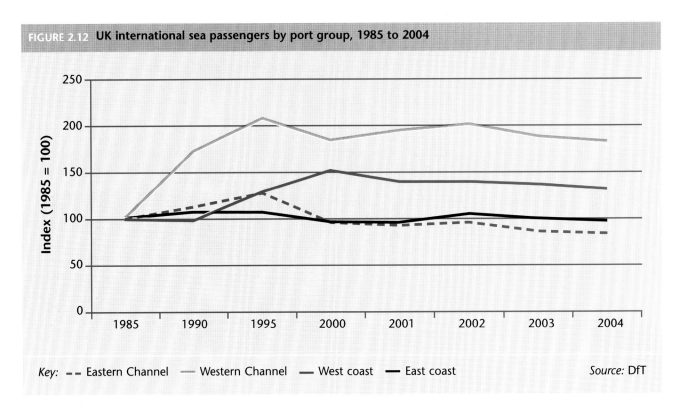

FIGURE 2.12 UK international sea passengers by port group, 1985 to 2004

Key: -- Eastern Channel — Western Channel — West coast — East coast

Source: DfT

east coast services to Germany, the Netherlands and Scandinavia. The Scottish islands and the Isle of Wight are served by additional domestic services.

In 2004, UK ports handled 26 million international ferry passengers and a further 23 million domestic passengers (excluding river ferries). Dover was the leading ferry port, handling about 55 per cent of international ferry passengers. Since 1975, international passengers have increased by 70 per cent, reaching a peak of 36 million in 1997 (Table 2.13).

Since 2000, passenger numbers have increased marginally on east coast routes (up 1 per cent) but have fallen on west coast routes (down 14 per cent), eastern Channel routes (down 13 per cent) and western Channel routes (down 1 per cent). Ferry passenger services have faced strong competition from low-cost airlines in recent years and were also affected by the abolition of duty-free allowances (Figure 2.12).

There is a strong connection between ro-ro freight and passenger services, as there are economic benefits from carrying both where this is feasible. As a result, the major international passenger ferry services operate from ports which also provide ro-ro freight facilities.

TABLE 2.14 Visits to and from the UK, 2004		
	Million visits	
	UK residents' visits abroad	**Overseas residents' visits to UK**
Air	50.4	20.0
Sea	9.0	4.8
Channel Tunnel	4.8	3.0
Total	64.2	27.8
Share by sea	**14%**	**17%**
Source: International Passenger Survey		

The International Passenger Survey (IPS) provides additional information about trends in personal travel. In 2004, around 14 per cent of visits by UK residents abroad and 17 per cent of overseas residents' visits to the UK were made by sea (Table 2.14). Visits abroad by UK residents by sea grew from 6.8 million in 1980 to 11 million in

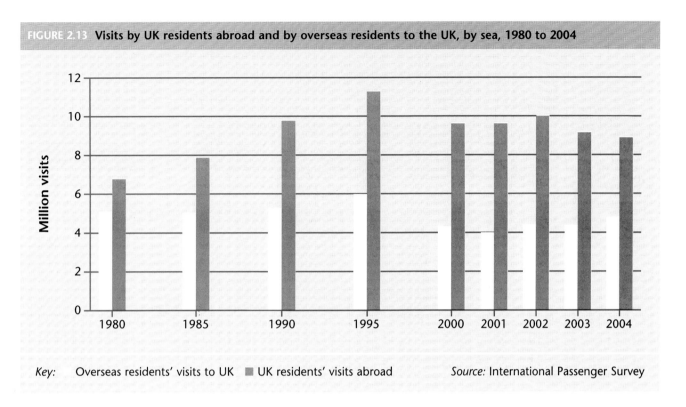

FIGURE 2.13 Visits by UK residents abroad and by overseas residents to the UK, by sea, 1980 to 2004

Key: Overseas residents' visits to UK   ■ UK residents' visits abroad   *Source:* International Passenger Survey

1995, falling back to 9.0 million in 2004. By contrast, overseas residents' visits to the UK grew more slowly, from 5.1 million to 6.0 million, falling back slightly to 4.8 million over the same period (Figure 2.13).

Domestic sea passengers

Domestic sea passenger trips cover sea crossings between mainland Britain and Northern Ireland, the Isle of Man, the Channel Islands, the Orkney and Shetland Isles and other inter-island routes (for example the Isles of Scilly , the Isle of Wight, the west of Scotland and services within Orkney and Shetland). In addition to these, there are trips by river ferries such as those on the Thames and Mersey.

There were 23 million domestic sea and inter-island passenger journeys in 2004. Passengers on sea crossings totalled 4.3 million, a 28 per cent increase since 1990. The busiest sea routes were between Great Britain and Northern Ireland (60 per cent of all sea passengers), then the Isle of Man (14 per cent), the Channel Islands (11 per cent) and Orkney and Shetlands (8 per cent). Since 2000, passenger journeys from Great Britain have risen on routes to Orkney and Shetlands (42 per cent), the Channel Islands (15 per cent) and the Isle of Man (10 per cent). Journeys to Northern Ireland have fallen by 9 per cent over the same period (Table 2.15).

Accompanied passenger vehicles

Many passengers passing through UK ports travel in their own cars or by coach. In 2004, 6.8 million accompanied passenger cars and 0.20 million coaches travelled through UK ports (Table 2.16). The most popular foreign route was France (3.7 million cars), whilst the busiest domestic route was Northern Ireland (1.1 million cars). Dover was the leading port, with 2.6 million passenger vehicles in 2004, 38 per cent of all vehicles (Figure 2.14).

TABLE 2.15 **Domestic waterborne passengers by route, 1990 to 2004**

	Waterborne passengers (000s)						
	1990	1995	2000	2001	2002	2003	2004
Sea crossings	3,382	3,883	4,273	4,094	4,314	4,436	4,318
GB–Northern Ireland	1,782	2,400	2,840	2,661	2,587	2,657	2,576
of which:							
Cairnryan–Larne	–	551	644	604	651	599	595
Liverpool–Belfast	–	58	154	130	137	150	158
Stranraer–Belfast	–	560	1,458	1,358	1,296	1,363	1,319
Stranraer–Larne	–	1,231	–	–	–	–	–
Troon–Belfast	–	–	364	362	332	368	303
GB–Isle of Man	–	415	531	496	572	590	586
of which							
Heysham–Douglas	–	260	240	224	252	261	272
Liverpool–Douglas	–	133	286	273	314	324	310
GB–Channel Islands	–	453	401	430	548	524	463
of which							
Poole–Jersey/Guernsey	–	–	204	175	279	263	227
Weymouth–Jersey/Guernsey	–	453	149	205	207	208	186
GB–Orkneys and Shetlands	–	221	239	208	250	291	339
Inter-island	15,976	16,857	16,662	17,120	17,658	18,398	18,627
of which:							
Hampshire–Isle of Wight	–	7,608	8,585	8,942	9,025	9,289	9,319
All Scottish routes[1]	–	8,049	6,840	6,942	7,094	7,562	7,762
All sea and inter-island journeys	19,358	20,740	20,935	21,214	21,972	22,851	22,945
River ferries	17,362	17,552	17,282	18,474	17,877	18,537	18,266
Total	**36,721**	**38,292**	**38,217**	**39,687**	**39,850**	**41,388**	**41,211**

1 Including Caledonian MacBrayne Clyde and Hebridean routes and routes within Orkney and Shetlands

Source: DfT

Trade vehicle imports and exports

The import and export of trade vehicles has risen two and half times since 1985, totalling 4.0 million vehicles in 2004. During this period there has been a marked concentration of the trade – 69 per cent of traffic took place at Southampton, London, Grimsby & Immingham and Bristol in 2004 (Table 2.17). Ports such as Harwich and Dover now have little or no traffic of this type. Whilst inward traffic has doubled since 1985 to 2.5 million vehicles, outward traffic has risen more than fourfold to 1.5 million. Coastwise movements represented about 4 per cent of inward and 6 per cent of outward traffic in 2004 (Figure 2.15).

The movement of motor vehicles by sea requires the availability of extensive quayside parking. The busiest UK port in 2004 was Southampton, with 742,000 vehicles (19 per cent of all traffic). Outward vehicle traffic outnumbered inward traffic at Southampton and also at Tyne ports, whilst the opposite was true at the other leading ports (Table 2.17). The majority of imported vehicles originated at ports in Europe (81 per cent) and the Far East (17 per cent), whilst the majority of exported vehicles were destined for ports in Europe (79 per cent) and North America (12 per cent).

TABLE 2.16 Accompanied passenger vehicles, 1990 to 2004

	1990	1995	2000	2001	2002	2003	2004
Passenger cars							
Dover	2,070	2,731	2,433	2,396	2,466	2,418	2,507
Portsmouth	742	824	934	976	1,011	915	891
Holyhead	302	408	500	464	488	501	481
Belfast	38	149	437	397	400	403	406
Stranraer	239	421	270	248	257	239	275
Harwich	233	267	285	272	280	254	244
Plymouth	161	169	175	176	192	187	189
Poole	200	133	176	200	234	216	186
Larne	332	441	155	149	164	175	174
Hull	166	169	217	197	186	167	165
Liverpool	51	32	37	133	148	162	162
Fishguard	169	130	194	180	183	157	156
Cairnryan	88	155	151	140	153	139	137
Tyne	31	56	73	63	121	123	113
Milford Haven	–	77	130	114	117	118	111
Newhaven	135	161	73	76	78	90	91
Heysham	58	50	123	97	86	75	76
Orkney	38	42	50	40	49	62	64
Forth	–	–	–	–	28	43	44
Swansea	26	51	41	38	41	41	40
Other ports	744	1,011	351	275	258	319	305
All cars	**5,821**	**7,477**	**6,806**	**6,631**	**6,939**	**6,804**	**6,818**
All buses and coaches	**218**	**234**	**203**	**194**	**201**	**188**	**199**

Source: DfT

TABLE 2.17 Trade vehicles through UK ports, 1985 to 2004[1]

	1985		1990		1995		2000		2004	
	Inwards	Outwards	Inwards	Outwards	Inwards	Outwards	Inwards	Outwards	Inwards	Outwards
Southampton	53	28	110	151	103	292	221	289	253	490
London	49	1	14	2	158	179	353	256	484	186
Grimsby & Immingham	101	1	137	8	169	71	279	145	469	191
Bristol	148	13	106	22	225	37	364	104	505	137
Tyne	–	–	–	–	59	175	23	191	165	291
Medway	183	9	208	77	242	22	325	53	368	57
Tees & Hartlepool	129	2	177	56	84	1	134	9	141	2
Belfast	–	–	51	3	81	5	98	22	61	6
Liverpool	4	35	5	21	3	34	21	48	16	40
Heysham	1	18	1	37	2	59	8	73	3	41
Other ports	509	205	510	227	197	101	28	50	27	20
Total	**1,177**	**312**	**1,319**	**603**	**1,324**	**977**	**1,853**	**1,242**	**2,492**	**1,461**

1 Major ports from 1995

Source: DfT

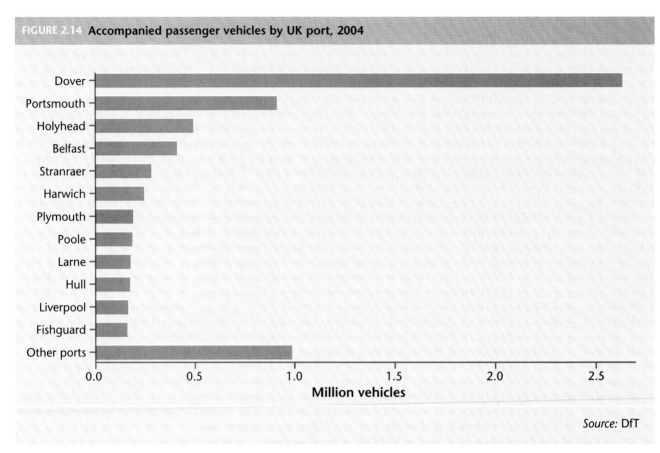

FIGURE 2.14 Accompanied passenger vehicles by UK port, 2004

Source: DfT

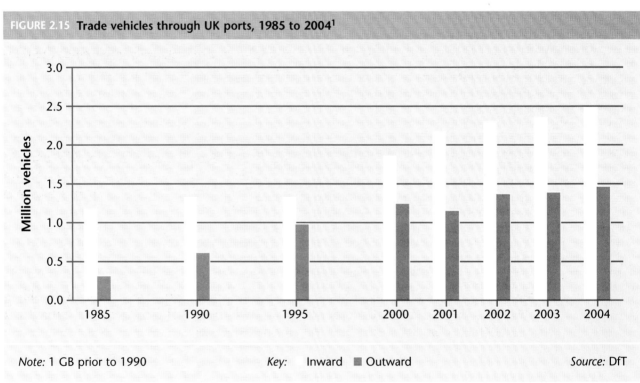

FIGURE 2.15 Trade vehicles through UK ports, 1985 to 2004[1]

Note: 1 GB prior to 1990   *Key:* Inward ■ Outward   *Source:* DfT

General cargo

General cargo consists of forestry products, iron and steel products and other general cargo including small containers. In comparison to other cargo types, this kind of traffic tends to be handled by a large number of UK ports, rather than being concentrated at a

TABLE 2.18 General cargo at UK ports, 2004				
	Thousand tonnes			
	Forest products	Iron and steel products	Other general cargo	Total
London	1,999	522	968	3,489
Medway	1,699	333	600	2,632
Newport	184	1,990	13	2,187
Grimsby & Immingham	525	810	436	1,771
Tees & Hartlepool	11	1,339	166	1,515
Hull	1,043	348	116	1,507
River Trent	176	1,298	2	1,476
Goole	389	684	352	1,424
Aberdeen	198	379	710	1,287
Liverpool	450	39	428	917
Forth	387	135	247	769
Clydeport	251	2	482	736
Felixstowe	696	–	26	722
Portsmouth	10	–	706	716
Cardiff	135	354	107	596
Other ports	2,405	1,824	1,368	5,597
All major ports	**10,558**	**10,057**	**6,726**	**27,341**

Source: DfT

few specialised ports (Table 2.18). General cargo has been relatively stable since the mid-1980s, rising from 26 million tonnes in 1985 to 27 million tonnes in 2004.

Forestry products

Forestry products accounted for 39 per cent of all general cargo at major UK ports in 2004 (11 million tonnes). London (2.0 million tonnes) and Medway (1.7 million tonnes) were the busiest ports for this type of cargo. Around 95 per cent of traffic was foreign imports in 2004.

Iron and steel products

Iron and steel products accounted for 37 per cent of all general cargo in 2004 (10 million tonnes). Newport (2.0 million tonnes), Tees & Hartlepool (1.3 million) and River Trent (1.3 million) were the busiest ports for this type of cargo.

Other general cargo

Other general cargo traffic includes miscellaneous cargo loaded in bundles, pallets and small containers. It accounted for one-quarter of all general cargo in 2004 (6.7 million tonnes) and is handled by most UK ports. This category also included 1.0 million tonnes of non-oil and gas traffic with offshore installations.

Ports handling cruise passengers

Cruise statistics are collected on the number of cruise journeys starting or finishing at a UK port. They exclude passengers flying out to join cruises which sail entirely in foreign waters.

The UK cruise market has grown considerably in recent years, rising from 207,000 passengers in 1995 to 767,000 in 2004 (Figure 2.16). The leading UK ports in 2004 were Southampton (62 per cent), Dover (20 per cent) and Harwich (12 per cent). Around 4 per cent of passengers passed through Falmouth and London (Figure 2.17). Since the mid-1990s there has been a sharp increase in passengers at the leading cruise ports and a fall in passengers passing through London.

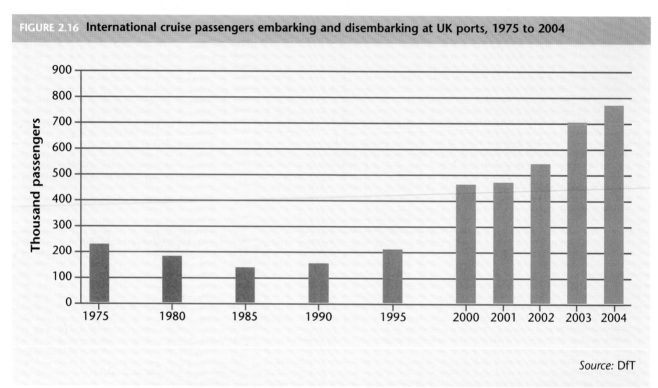

FIGURE 2.16 International cruise passengers embarking and disembarking at UK ports, 1975 to 2004

Source: DfT

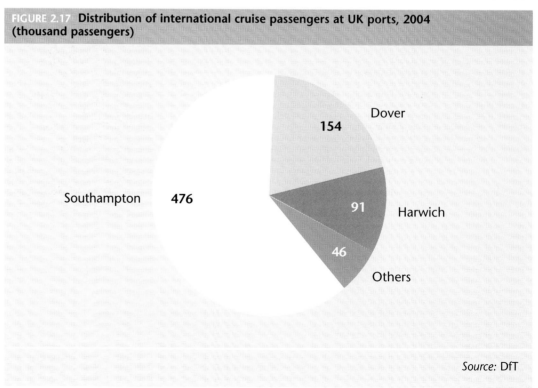

FIGURE 2.17 Distribution of international cruise passengers at UK ports, 2004 (thousand passengers)

Source: DfT

Fishing ports

There are many harbours in the UK with facilities for commercial fishing. There are 226 fishing ports in Scotland, 211 in England, 42 in Northern Ireland and 36 in Wales.

UK fishing ports handled £516 million and 584 million tonnes of fish landed by UK and foreign vessels in 2004 (Table 2.19). Around a fifth of the total value was landed by foreign registered vessels. Total tonnage landed has risen steadily since 2000, although landings are currently only around three-quarters of landings in 1995. The principal ports are in Scotland and south-west England, namely Peterhead (16 per cent of the total value landed by UK vessels in 2004), Fraserburgh (9 per cent), Lerwick (8 per cent), Brixham (4 per cent) and Newlyn (4 per cent) (Figures 2.18 and 2.19).

The fishing industry has undergone major changes in recent years. Boundary disputes in the 1970s led to the introduction of territorial fishing limits around Iceland, Greenland, Canada and Russia, which severely impacted the UK's distant water fleets. In addition, fishing quotas and fleet reduction targets have been introduced by the European Union.

The UK fishing fleet consists of around 7,000 registered fishing vessels, providing almost 12,000 full and part-time fishing jobs. In addition, there are also many shore-based jobs supporting the fish-processing industry. *UK Sea Fisheries Statistics* estimates that the fisheries sector contributes half a billion pounds to UK GDP and employs around 18,000 people in the wider fish-processing sector.

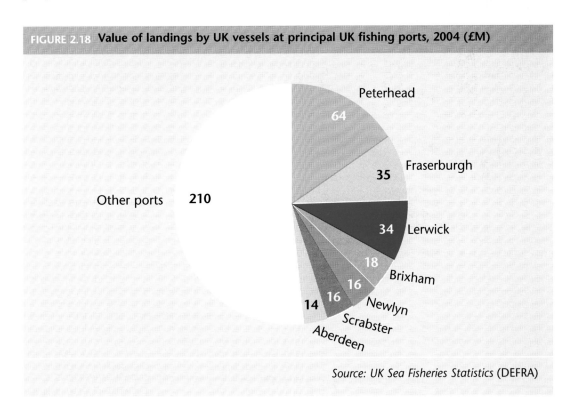

FIGURE 2.18 Value of landings by UK vessels at principal UK fishing ports, 2004 (£M)

Source: UK Sea Fisheries Statistics (DEFRA)

FIGURE 2.19 Fishing ports landing £2 million or more of fish, by UK vessels, 2004

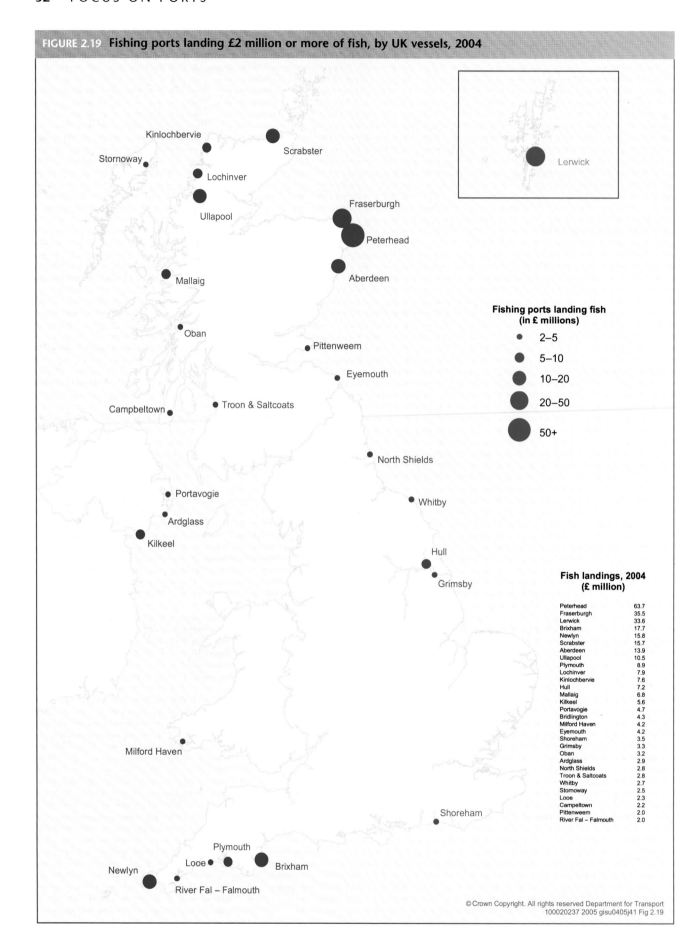

Fishing ports landing fish (in £ millions)

- 2–5
- 5–10
- 10–20
- 20–50
- 50+

Fish landings, 2004 (£ million)

Peterhead	63.7
Fraserburgh	35.5
Lerwick	33.6
Brixham	17.7
Newlyn	15.8
Scrabster	15.7
Aberdeen	13.9
Ullapool	10.5
Plymouth	8.9
Lochinver	7.9
Kinlochbervie	7.6
Hull	7.2
Mallaig	6.8
Kilkeel	5.6
Portavogie	4.7
Bridlington	4.3
Milford Haven	4.2
Eyemouth	4.2
Shoreham	3.5
Grimsby	3.3
Oban	3.2
Ardglass	2.9
North Shields	2.8
Troon & Saltcoats	2.8
Whitby	2.7
Stornoway	2.5
Looe	2.3
Campeltown	2.2
Pittenweem	2.0
River Fal – Falmouth	2.0

TABLE 2.19 Value and tonnage of fish landings by UK and foreign vessels at UK ports, 1994 to 2004		
	£M	**Million tonnes**
1994	500	751
1995	521	791
1996	544	711
1997	511	644
1998	543	614
1999	522	560
2000	483	526
2001	488	530
2002	490	538
2003	490	557
2004	516	584
Source: UK Sea Fisheries Statistics (DEFRA)		

Inland waters traffic

Non-sea-going inland waters traffic totalled 2.6 million tonnes lifted in 2004. This represented 6 per cent of all inland waters traffic and 2 per cent of all UK domestic waterborne freight. More than half of UK non-sea-going traffic is lifted on the Thames.

Freight facility grants are available to help meet any extra costs associated with the transfer of freight away from the congested road network. A company wishing to move freight within Great Britain by water or rail can apply for a grant towards the capital expenditure involved. The total grant awarded is dependent on the value of the environmental benefits and an 'assessment' of the financial case for the grant. Table 2.20 provides information about recent inland waters and coastal freight grants. The large figure for lorry miles saved in 2001 relates to port development at Rosyth in respect of the Zeebrugge service.

Ports and recreation

Recreational sailing has grown considerably in recent years and is now a substantial element for some ports. These developments have occurred as an alternative to other commercial activities and some ports are dependent upon them to finance their conservancy and regulation. The British Marine Federation estimated that there were around 450,000 recreational boats in the UK in 2004 (defined as craft over 2.5 m). Examples of large concentrations of leisure craft can be seen on the Solent, the south-west of England and the west coast of Scotland.

TABLE 2.20 Freight facilities grants awarded for inland waterway, coastal and short sea shipping projects, 1997 to 2004[1]	1997	1998	1999	2000	2001	2002	2003	2004
Inland waterways projects								
Number of grants[2]	1	3	8	7	7	6	4	2
Value of grants (£ million)	0.1	1.4	7.5	4.5	3.6	1.7	1.1	4.7
Lorry miles saved (million)[3]	0.02	0.11	0.49	0.28	12.74	2.70	2.93	6.69
Coastal/short sea shipping projects								
Number of grants[2]	–	–	–	2	3	4	3	–
Value of grants (£ million)	–	–	–	5.1	11.4	3.9	1.5	–
Lorry miles saved (million)[3]	–	–	–	11.8	1,309.14	7.99	10.42	–
All projects								
Number of grants[2]	1	3	8	9	10	10	7	2
Value of grants (£ million)	0.1	1.4	7.5	9.6	15.0	5.6	2.7	4.7
Lorry miles saved (million)[3]	0.02	0.11	0.49	12.08	1,321.88	10.70	13.35	6.69

1 Section 140 of Railways Act 1993 and Section 272 of Transport Act 2000; 2 All grants by calendar year; 3 Lorry miles saved are those over the lifetime of the grant totalled in the year it was awarded

Source: DfT and Scottish Executive

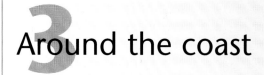

Around the coast

Introduction

This chapter looks at ports in a regional context, using the port groupings shown in Table 3.1. These are standard groupings for ports analyses and do not correspond to government office regions. Both freight and passenger traffic is covered in the chapter and where appropriate fish landings. (Unless stated otherwise, values and tonnage of fish landings refer to UK vessels). It should be noted that some ports (for example Grimsby & Immingham, and Tees & Hartlepool) are treated as single ports in the statistics. Also, where reference is made to Forth Ports, this refers to the port location not the company.

In 2004, the most important regions, in tonnage terms, were Thames and Kent, with 91 million tonnes (16 per cent of UK port tonnage), the east coast of Scotland 88 million tonnes (15 per cent) and the Humber 84 million tonnes (also 15 per cent). The smallest amounts of port traffic were handled in the West Country and in Wash & northern East Anglia, each of these regions handling 1 per cent of UK port traffic (Table 3.2).

Total UK traffic has grown by 149 million tonnes (35 per cent) since 1980, but this growth has not been evenly distributed across the regions. The largest growth, 50 million tonnes (145 per cent), was in the Humber region. Traffic at Haven ports increased by 19 million tonnes (143 per cent), reflecting Felixstowe's growth in container traffic. Ports on the west and north coast of Wales have experienced the largest fall in tonnage since 1980; the fall of 5 million tonnes (9 per cent) was largely due to the closure of Anglesey Marine Terminal in 1988 (Table 3.2).

TABLE 3.1 Geographical port groupings	
Port group region	**Geographical definition**
Thames and Kent	Ports in South Essex, Greater London and Kent (plus Rye, Sussex)
Sussex and Hampshire	Ports in Sussex (excl. Rye), Hampshire and the Isle of Wight
West Country	Ports in Dorset, South Devon and Cornwall
Bristol Channel	Ports in North Devon, Somerset, Gloucestershire, Avon and South Wales (west to Llanelli)
Wales (west and north coasts)	Ports in Wales, west and north of Llanelli
Lancashire and Cumbria	Ports in Merseyside, Greater Manchester, Lancashire and Cumbria
Scotland (west coast)	Ports in West Scotland, including the Western Isles, to the midpoint on the north coast of Scotland
Scotland (east coast)	Ports in East Scotland, including Orkneys and Shetlands, from the midpoint on the north coast of Scotland (including Scrabster)
North-east	Ports in Northumberland, Durham and North Yorkshire
Humber	Ports in East Yorkshire and North Lincolnshire (incl. ports on the Rivers Hull, Humber, Trent and Ouse)
Wash and northern East Anglia	Ports in South Lincolnshire, Cambridgeshire, Norfolk and North Suffolk
Haven	Ports in South Suffolk and North Essex (incl. Felixstowe, Harwich and Ipswich)
Northern Ireland	Ports in Northern Ireland

TABLE 3.2 Freight traffic at all UK ports by port group, 1980 to 2004

					Million tonnes				
	1980	1985	1990	1995	2000	2001	2002	2003	2004
Thames and Kent	80.0	75.3	90.1	84.5	83.0	87.0	88.6	87.9	90.9
Sussex and Hampshire	30.1	32.6	37.1	40.9	42.7	43.9	42.4	43.9	47.0
West Country	7.0	7.2	7.7	7.3	6.7	7.2	7.0	7.1	6.9
Bristol Channel	20.4	23.9	27.3	28.9	29.7	28.1	23.2	26.8	27.6
Wales (west and north coasts)	48.5	42.9	35.0	36.3	38.7	38.4	39.7	38.0	44.0
Lancashire and Cumbria	31.3	25.6	36.5	44.2	45.0	45.0	43.3	44.7	45.2
Scotland (west coast)	11.8	13.4	16.6	17.2	17.6	21.0	20.1	19.3	22.0
Scotland (east coast)	79.7	115.3	81.0	109.7	112.9	102.8	102.1	91.2	88.4
North-east	52.1	43.5	51.8	53.4	56.4	55.7	55.2	59.1	59.3
Humber	34.3	46.0	59.7	69.0	77.7	78.5	79.7	80.9	84.0
Wash and northern East Anglia	4.0	5.7	6.8	5.3	4.5	3.5	3.5	4.0	2.9
Haven	12.9	18.0	25.7	31.4	36.8	34.1	32.1	30.6	31.4
Northern Ireland	12.1	13.5	16.7	20.3	21.4	21.2	21.4	22.0	23.4
Total	**424.1**	**462.9**	**492.0**	**548.2**	**573.0**	**566.4**	**558.3**	**555.7**	**573.1**

Source: DfT

TABLE 3.3 UK major port traffic by port group, 2004

	Total traffic Million tonnes	Share of total (%)	Of which: liquid bulk Million tonnes	Liquid bulk as % of total traffic
Thames and Kent	90.3	16	21.6	24
Sussex and Hampshire	46.0	8	27.2	59
West Country	5.3	1	1.4	27
Bristol Channel	26.0	5	3.9	15
Wales (west and north coasts)	42.9	8	37.5	87
Lancashire and Cumbria	44.1	8	17.4	39
Scotland (west coast)	20.8	4	3.5	17
Scotland (east coast)	85.6	15	78.2	91
North-east	57.9	10	37.0	64
Humber	83.8	15	35.6	42
Wash and northern East Anglia	1.3	–	0.2	12
Haven	31.2	6	0.6	2
Northern Ireland	23.0	4	3.4	15
Total	**558.2**	**100**	**267.4**	**48**

Source: DfT

Liquid bulk traffic makes up a large percentage of UK major port traffic (48 per cent in 2004) and is particularly important at ports on Scotland's east coast (91 per cent of all traffic) and in North and West Wales (87 per cent) – see Table 3.3.

Thames and Kent

The Thames and Kent region handled 91 million tonnes (16 per cent) of UK port traffic in 2004 – the largest regional total. The three primary ports in the region are London (59 per cent of total traffic), Dover (23 per cent) and Medway (16 per cent). London's traffic has fallen by 18 per cent since 1965, falling sharply in the early 1970s before levelling out around the current tonnage. Dover has experienced almost continuous growth since

TABLE 3.4 Freight at Thames and Kent ports, 1965 to 2004												
						Million tonnes						
	1965	1970	1975	1980	1985	1990	1995	2000	2001	2002	2003	2004
London	64.6	63.8	50.3	54.2	51.6	58.1	51.4	47.9	50.7	51.2	51.0	53.3
Dover	1.4	2.1	3.7	6.7	9.3	13.0	12.7	17.4	19.1	20.2	18.8	20.8
Medway	22.3	26.9	21.7	17.2	10.4	13.6	14.2	15.3	14.9	14.8	15.6	14.5
Ramsgate	–	0.1	–	0.1	1.3	2.9	4.8	1.2	1.4	1.8	1.8	1.7
Wallasea	..	..	–	0.1	0.1	0.1	0.2	0.1	0.1	0.2	0.2	0.2
Folkestone	–	..	0.4	0.4	0.5	0.7	0.1	0.6	0.3	–	0.1	0.1
Whitstable	..	0.1	0.2	0.2	0.2	0.3	0.2	0.2	0.2	0.2	0.1	0.1
Brightlingsea	..	..	..	0.1	0.5	0.4	0.1	–	0.2	0.1	0.1	0.1
Colchester	0.6	0.7	0.7	1.0	1.2	0.7	0.5	0.2	–	–	–	–
Other ports	..	..	..	0.1	0.3	0.3	0.4	–	0.1	0.1	0.1	0.1
Total[1]	89.0	94.0	77.5	80.0	75.3	90.1	84.5	83.0	87.0	88.6	87.9	90.9

1 Estimated for earlier years where figures for some ports not available

Source: DfT

1965. Traffic at Medway more than halved between 1970 and 1985, but since then has grown by half. Traffic at other ports in the region made up only 2.5 per cent of the total in 2004 (Table 3.4).

London

The Port of London stretches from Teddington Lock in West London to the North Sea, a distance of 95 miles (150 kilometres). Once the biggest port in the world, London is still one of the top three UK ports, although it has changed substantially over the years in response to changing demands of the industry. The enclosed docks and many of the upriver wharves closed during the 1970s and 1980s as activity moved further down river. Of its numerous operational cargo terminals, the largest are Tilbury, Purfleet, Thames Europort (Dartford) and Coryton oil refinery. At Silvertown there is one of the largest cane sugar refineries in the world.

TABLE 3.5 Port traffic at London, 2004					
			Million tonnes		
	Imports	Exports	Total foreign traffic	Domestic	Total
Liquid bulk	12.0	2.1	14.1	6.2	20.3
of which: crude oil	3.9	–	3.9	3.7	7.7
oil products	7.4	2.0	9.4	2.1	11.5
Dry bulk	5.5	1.1	6.6	7.6	14.2
Containers	6.4	2.7	9.1	–	9.1
Mobile self-propelled vehicles	0.7	0.3	1.0	–	1.0
Mobile non-self-propelled vehicles	3.5	1.6	5.1	–	5.1
Other cargo	3.1	0.3	3.4	–	3.5
Total	31.2	8.2	39.4	13.9	53.3

Source: DfT

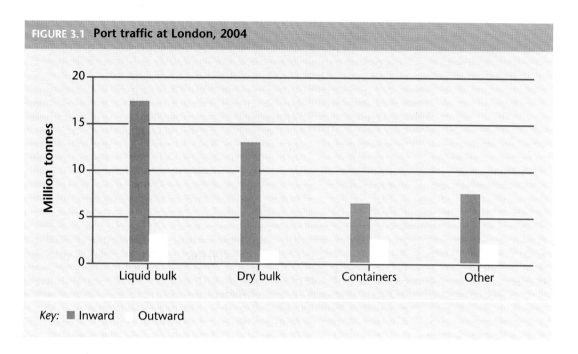

FIGURE 3.1 Port traffic at London, 2004

Key: ■ Inward □ Outward

Of the 53 million tonnes handled in 2004, the leading cargoes were oil products (22 per cent), miscellaneous dry bulk (18 per cent) and containers (17 per cent). London was the UK's third busiest port in terms of TEU in 2004 (979,000). Crude oil and unaccompanied trailers are also important (Table 3.5 and Figure 3.1).

Internal freight and ferry passengers are also carried on the Thames. Internal freight traffic declined sharply between 1985 and 1990 from 3.5 million to 2.0 million tonnes. Traffic fell again in 2004 from 2.0 million to 1.5 million tonnes (Figure 3.2).

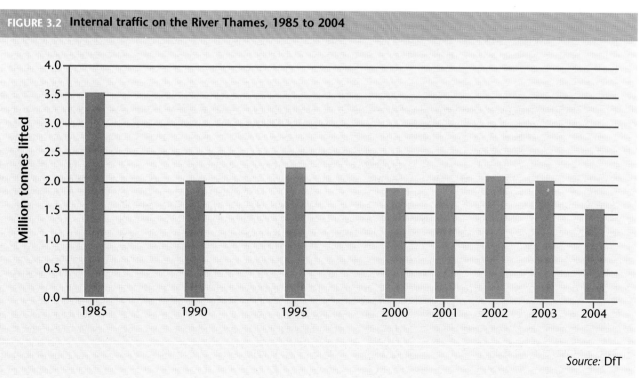

FIGURE 3.2 Internal traffic on the River Thames, 1985 to 2004

Source: DfT

Dover

Dover is the UK's principal ferry port. In 2004, it handled 21 million tonnes, of which over 95 per cent was ro-ro traffic. There were 25,000 ro-ro vessel arrivals, carrying 14 million passengers, mainly to and from Calais. Dover is also an important cruise port, with more than 150,000 passengers passing through the cruise terminal in 2004.

Medway

The port of Medway includes the ports of Kingsnorth, Thamesport, the Isle of Grain and Sheerness. Medway handled 15 million tonnes of cargo in 2004, including containers (26 per cent), coal (23 per cent) and miscellaneous dry bulk (20 per cent). There has been a steady growth in traffic since the low levels in the mid-1980s, including recent growth in containerised traffic (632,000 TEU in 2004 – the fourth largest in the UK). Medway was also the second busiest UK port for forestry products, handling 16 per cent of the UK total.

Other ports

The remaining Thames and Kent ports handled 2.3 million tonnes (3 per cent) in 2004. Ramsgate (1.7 million tonnes of mainly ro-ro traffic) was the busiest of these. Traffic at the port has experienced growth in recent years, although the current level is lower than the peak of 4.8 million tonnes in 1995.

Sussex and Hampshire

The Sussex and Hampshire region handled 47 million tonnes of cargo in 2004, representing 8 per cent of total UK port traffic. The region's traffic has grown by more than half since 1980. Southampton was by far the busiest port in 2004 (82 per cent of total traffic), followed by Portsmouth (11 per cent) (Table 3.6). The region also has ferry links to the Isle of Wight (9.3 million passengers in 2004).

TABLE 3.6   Freight at Sussex and Hampshire ports, 1965 to 2004

	1965	1970	1975	1980	1985	1990	1995	2000	2001	2002	2003	2004
						Million tonnes						
Southampton	24.4	28.2	25.3	23.9	25.2	28.8	32.4	34.8	35.7	34.2	35.8	38.4
Portsmouth	0.9	1.2	1.1	1.4	2.0	2.6	4.4	4.5	4.3	4.4	4.2	4.9
Shoreham	2.9	3.0	2.2	2.9	2.7	2.6	2.0	1.8	1.8	1.8	1.7	1.7
Newhaven	0.4	0.6	1.3	1.2	1.6	2.0	1.0	0.6	1.0	0.9	0.9	0.9
Cowes (IoW)	..	0.2	0.1	0.3	0.3	0.4	0.3	0.4	0.5	0.6	0.7	0.5
Littlehampton	0.1	0.2	0.4	0.3	0.4	0.3	0.2	0.2	0.2	0.2	0.2	0.1
Other ports	..	..	..	0.2	0.5	0.4	0.5	0.4	0.5	0.4	0.4	0.4
Total[1]	29.0	33.5	30.5	30.1	32.6	37.1	40.9	42.7	43.9	42.4	43.9	47.0

1 Estimated for earlier years where figures for some ports not available

Source: DfT

Southampton

Southampton has seen generally consistent traffic growth since 1980. It was the fifth largest UK port in 2004, handling 38 million tonnes (7 per cent of UK traffic). Crude oil made up 43 per cent of total traffic, oil products and liquefied gas a further 27 per cent and container traffic 20 per cent (Table 3.7). Southampton was the second largest UK container port in 2004 (1.4 million TEU), with about 80 per cent of this traffic being with the Far East.

Southampton also has the largest share of the UK cruise passenger market (62 per cent of UK cruise passengers in 2004) and the largest cruise ships call there. The port has three cruise terminals and is home to the *QE2* and the recently commissioned *Queen Mary 2*.

Portsmouth

Portsmouth handled 4.9 million tonnes in 2004, 70 per cent of which was ro-ro traffic to France. Traffic has risen more than threefold since 1980, including particularly strong growth between 1990 and 1995. Portsmouth was the second largest UK passenger ferry port in 2004, with 3.1 million passengers sailing to France, Spain and the Channel Islands. In addition, there are also services to the Isle of Wight. In 2004, fish landings at Portsmouth were valued at £1.8 million.

Shoreham

Shoreham handled 1.7 million tonnes in 2004, 67 per cent of which was miscellaneous dry bulk and 11 per cent oil products. Freight traffic has levelled out in recent years after falling by more than 40 per cent since 1970. Fish landings were valued at £3.5 million in 2004.

Newhaven

Total traffic at Newhaven was 0.9 million tonnes in 2004. Whilst this was less than half the level in 1990, traffic has recently increased since the low levels of the late 1990s. Ro-ro traffic and sea-dredged aggregates formed the main types of traffic in 2004. There were also 360,000 ferry passenger movements to and from Dieppe.

TABLE 3.7   Port traffic at Southampton, 2004

	Million tonnes				
	Imports	Exports	Total	Domestic	Total
Liquid bulk	11.8	6.1	17.9	9.1	27.0
of which: crude oil	8.4	0.5	9.0	7.7	16.7
Dry bulk	0.4	0.4	0.8	1.2	2.0
Containers	4.2	3.5	7.7	0.1	7.8
Mobile self-propelled vehicles	0.4	1.0	1.4	–	1.4
Mobile non-self-propelled vehicles	–	0.1	0.1	–	0.1
Other cargo	–	0.2	0.2	–	0.2
Total	**16.9**	**11.2**	**28.1**	**10.3**	**38.4**

Source: DfT

Other ports

Other ports in the Hampshire and Sussex region include Cowes (0.5 million tonnes in 2004) and Littlehampton (0.1 million). In addition, over 9 million passengers travelled to and from the Isle of Wight. These services run between Lymington, Portsmouth and Southampton on the mainland and Cowes, Fishbourne, Ryde and Yarmouth on the Isle of Wight.

West Country

The West Country handled 6.9 million tonnes in 2004 (1.2 per cent of UK traffic). Traffic grew steadily between 1965 and the late 1980s before falling back marginally. Plymouth, Poole and Fowey are the leading ports, making up three-quarters of total traffic. Ports such as Fowey and Par are important for china clay exports (Table 3.8). The fishing industry is also of economic importance to the region, in particular Brixham, Newlyn and Plymouth (Table 3.9).

TABLE 3.8 Freight at West Country ports, 1965 to 2004

	Million tonnes											
	1965	1970	1975	1980	1985	1990	1995	2000	2001	2002	2003	2004
Plymouth	1.6	1.9	1.4	1.7	1.6	1.5	1.7	1.8	1.9	1.9	2.1	2.2
Poole	1.0	1.1	1.0	1.3	1.5	1.9	1.7	1.3	1.8	1.8	1.6	1.8
Fowey	0.6	0.9	1.0	1.2	1.5	1.6	1.7	1.5	1.5	1.5	1.4	1.3
Teignmouth	0.3	0.4	0.4	0.5	0.6	0.9	0.7	0.7	0.7	0.7	0.7	0.6
Falmouth	0.3	0.3	0.5	0.4	0.2	0.6	0.5	0.6	0.5	0.4	0.4	0.4
Par	0.9	1.1	0.6	0.8	0.6	0.8	0.7	0.6	0.5	0.5	0.3	0.3
Exmouth (incl. Exeter)	0.1	0.1	0.1	0.2	0.6	–	–	–	–	–	–	–
Other ports	..	..	..	0.9	0.6	0.5	0.4	0.3	0.4	0.4	0.5	0.4
Total¹	**5.3**	**6.3**	**5.5**	**7.0**	**7.2**	**7.7**	**7.3**	**6.7**	**7.2**	**7.0**	**7.1**	**6.9**

1 Estimated for earlier years where figures for some ports not available

Source: DfT

Plymouth

The port of Plymouth includes the commercial port area as well as the naval base at Devonport. Total traffic was 2.2 million tonnes in 2004, consisting mainly of oil products (57 per cent) and miscellaneous dry bulk (38 per cent). Traffic at Plymouth has grown by around 45 per cent since 1990. Ro-ro vessels also carried 617,000 ferry passengers and 112,000 tonnes of freight, whilst fish landings totalled £8.9 million.

TABLE 3.9 Fish landings by UK vessels at principle West Country fishing ports, 2004

	Tonnes	£ million
Brixham	10,818	17.6
Newlyn	7,634	15.8
Plymouth	8,984	8.9
Looe	1,521	2.3
River Fal & Falmouth	1,425	2.0
Weymouth	4,042	1.6
Exeter, Exmouth & Teignmouth	1,472	1.4
Padstow	560	1.3
Salcombe	975	1.2
Dartmouth, River Dart & Kingswear	969	1.2
Poole	520	1.1
Total	**38,920**	**52.3**

Source: UK Sea Fisheries Statistics (DEFRA)

Poole

Poole was the second busiest West Country port after Plymouth in 2004 (1.8 million tonnes). Steady growth occurred between 1975 and 1990, when traffic almost doubled to 1.9 million tonnes. Ro-ro traffic (58 per cent) and miscellaneous dry bulk (16 per cent) were the major cargoes in 2004. There were also 747,000 ferry passengers to France and the Channel Islands and fish landings valued at £1.1 million. Poole is also a major centre for leisure craft.

Fowey

Fowey handled 1.3 million tonnes in 2004. Exports of china clay formed the main cargo, three-quarters of which went to Scandinavia and Canada. Traffic at the port grew from 0.6 million to 1.7 million tonnes between 1965 and 1995.

Other ports

The remaining ports in the south-west region handled 1.7 million tonnes in 2004. These included Teignmouth (0.6 million), Falmouth (0.4 million) and Par (0.3 million, compared to 1.1 million tonnes in 1970). Falmouth's extensive deep-water moorings are used regularly for cruise calls as well as fuel bunkering. Weymouth handled 186,000 Channel Islands ferry passengers in 2004 and £1.6 million of fish landings. There is a ferry service between Penzance and the Isles of Scilly, which is important both for the local community and tourism.

Bristol Channel

The Bristol Channel region covers the area from North Devon through to Llanelli in South Wales. In 2004, it handled 28 million tonnes of traffic (5 per cent of UK traffic). The major share (70 per cent) was at Bristol and Port Talbot, with coal, ores and oil products making up the largest tonnage. The region has experienced peaks and troughs since 1965, reaching 31 million tonnes in 1970 before falling to 20 million tonnes in 1980 (Table 3.10).

Bristol

Bristol handled 11 million tonnes of traffic in 2004, with coal (34 per cent) and oil products (22 per cent) making up the largest tonnage. It was the fourth largest UK port for coal (9 per cent of total traffic), whilst the movement of trade vehicles is also important. Bristol's traffic fell to under 4 million tonnes in the 1980s, since when it has seen fairly steady growth (Figure 3.3). Port activity is mainly centred on Avonmouth and Portbury due to the large tidal range on the River Avon.

Port Talbot

Port Talbot handled 8.6 million tonnes of traffic in 2004. This consisted mainly of ores (61 per cent) and coal (34 per cent), which was largely imported. It was the third largest UK port for ores traffic in 2004 (29 per cent of total UK traffic). From a low of 2.4 million tonnes in 1980, traffic at Port Talbot had increased to almost 12 million tonnes by 2000.

TABLE 3.10 Freight at Bristol Channel ports, 1965 to 2004

| | | | | | | Million tonnes | | | | | | |
	1965	1970	1975	1980	1985	1990	1995	2000	2001	2002	2003	2004
Bristol	9.3	7.5	5.3	4.9	3.8	4.9	7.3	9.6	10.9	10.1	11.4	10.8
Port Talbot	3.7	3.7	3.7	2.4	6.8	8.9	11.0	11.7	8.3	5.0	7.8	8.6
Newport	4.5	3.7	3.5	2.0	2.5	3.2	2.5	2.7	3.0	3.1	2.8	3.4
Cardiff	2.7	3.7	3.2	2.6	2.0	2.5	2.4	2.7	2.7	2.2	2.3	2.5
Swansea	6.6	8.1	6.5	5.4	5.2	4.9	4.0	1.0	1.3	1.1	0.8	0.7
Gloucester & Sharpness	0.9	0.8	0.8	1.0	0.7	0.5	0.4	0.6	0.5	0.6	0.6	0.5
Barry	2.0	1.8	0.8	0.8	1.1	1.0	0.4	0.6	0.6	0.5	0.5	0.4
Neath	..	0.3	0.3	0.2	0.8	0.9	0.5	0.5	0.5	0.4	0.4	0.4
Bridgwater	0.4	0.6	0.4	0.5	0.4	0.2	0.1	0.1	0.1	0.1	0.1	0.1
Watchet	..	0.1	0.1	0.1	0.1	0.1	–	–	–	–	–	–
Penarth	0.3	0.4	0.2	0.2	0.1	–	–	–	–	–	–	–
Other ports	..	..	..	0.2	0.3	0.4	0.2	0.2	0.2	0.1	0.2	0.1
Total[1]	30.5	31.0	25.0	20.4	23.9	27.3	28.9	29.7	28.1	23.2	26.8	27.6

1 Estimated for earlier years where figures for some ports not available

Source: DfT

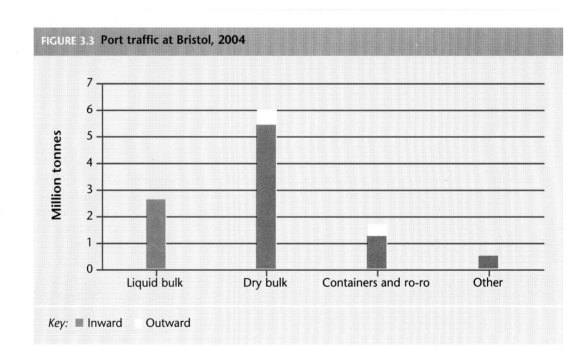

FIGURE 3.3 Port traffic at Bristol, 2004

Key: ■ Inward   □ Outward

Newport

Newport handled 3.4 million tonnes of traffic in 2004. Iron and steel products (58 per cent) and miscellaneous dry bulk (21 per cent) represented the largest tonnage. Newport was the UK's busiest port for iron and steel products in 2004, handling a fifth of all such traffic. Traffic fell from 4.5 million tonnes in 1965 to 2.0 million tonnes in 1980, before recovering to 3.2 million by 1990.

Cardiff

Cardiff handled 2.5 million tonnes in 2004 – almost half of which comprised oil products received from other UK ports. Traffic at Cardiff peaked in 1970 at 3.7 million tonnes, falling to 2.0 million by 1985.

Swansea

Swansea handled 0.7 million tonnes in 2004, half of which was dry bulk. Traffic halved from 8 million to 4 million tonnes between 1970 and 1995, before falling sharply to 1 million tonnes by 2000. The port handles a range of different cargoes and also has a ferry service to Cork (116,000 passengers in 2004).

Other ports

Other ports in the Bristol Channel region handled 1.6 million tonnes in 2004, including Gloucester & Sharpness (0.5 million), Barry (0.4 million) and Neath (0.4 million).

West and North Wales

Ports in West and North Wales handled 44 million tonnes of traffic in 2004 (8 per cent of UK traffic). Milford Haven is the dominant port, handling 86 per cent of the region's traffic. Holyhead is the other significant port and is also one of the UK's leading ferry ports (Table 3.11).

Milford Haven

The deep-water harbour of Milford Haven accommodates a number of oil refineries as well as Milford Docks and Pembroke, which is the base for ro-ro ferries to Rosslare. Total traffic of 39 million tonnes in 2004 was almost entirely oil products (55 per cent) and crude oil (41 per cent). It was the UK's busiest port for oil products in 2004, handling a quarter of all traffic (Figure 3.4). Port traffic has been fairly stable since the mid-1980s, having peaked at around 45 million tonnes in the mid-1970s. In 2004, 378,000 ferry

TABLE 3.11 Freight at Wales (west and north coast) ports, 1965 to 2004												
						Million tonnes						
	1965	1970	1975	1980	1985	1990	1995	2000	2001	2002	2003	2004
Milford Haven	24.8	41.3	44.9	39.3	32.4	32.2	32.5	33.8	33.8	34.5	32.7	38.5
Holyhead	0.3	0.3	0.8	1.2	1.2	1.2	2.3	3.4	3.2	3.3	3.3	3.9
Mostyn	0.1	0.2	0.1	0.1	0.2	0.3	0.1	0.3	0.3	0.9	0.9	0.7
Fishguard	0.1	0.1	0.2	0.3	0.3	0.4	0.5	0.4	0.3	0.4	0.5	0.5
Anglesey marine terminal	–	–	–	6.8	7.8	–	–	–	–	–	–	–
Caernarfon	0.1	0.1	0.1	0.1	0.1	–	–	–	–	–	–	–
Other ports	..	..	..	0.6	0.9	0.9	0.9	0.8	0.7	0.6	0.5	0.4
Total[1]	26.0	42.5	46.5	48.5	42.9	35.0	36.3	38.7	38.4	39.7	38.0	44.0

1 Estimated for earlier years where figures for some ports not available

Source: DfT

passengers passed through the port and fish landings were valued at more than £4 million.

Holyhead

Holyhead's port traffic has grown strongly in recent years, increasing from 1.2 million tonnes in 1990 to 3.9 million tonnes in 2004. More than 90 per cent of this was ro-ro traffic to Dublin and Dun Laoghaire, with the remainder being imported ores. Holyhead also handled 2.3 million ferry passengers in 2004 – the UK's third largest passenger ferry port.

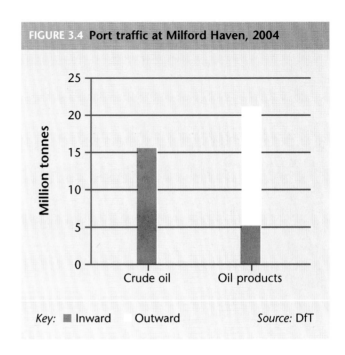

FIGURE 3.4 **Port traffic at Milford Haven, 2004**

Key: ■ Inward   Outward   Source: DfT

Other ports

Other ports in the region handled 1.6 million tonnes in 2004. These included Mostyn (0.7 million) and Fishguard (0.5 million). Fishguard also handled 614,000 ferry passengers, whilst the service from Mostyn closed in early 2004.

Lancashire and Cumbria

Lancashire and Cumbria handled 45 million tonnes in 2004 (8 per cent of UK traffic). The region has important ro-ro links to Ireland. Liverpool is the dominant port (71 per cent in 2004), followed by Manchester and Heysham. Traffic in the region more than halved between 1965 and 1985, reaching a low of 25 million tonnes. However, this had recovered to 44 million tonnes by 1995 (Table 3.12).

Liverpool

Liverpool handled 32 million tonnes in 2004, including crude oil (35 per cent), various dry bulks (27 per cent) and ro-ro traffic (18 per cent) (Table 3.13). The port, which includes the ports of Tranmere, Seaforth and Birkenhead, has seen significant growth in recent years, following the slump in traffic which took place during the late 1970s and 1980s. Traffic reached a low of 10 million tonnes in the late 1980s but had recovered to 30 million tonnes by 1995.

In 2004, Liverpool was the fifth largest UK container port (603,000 TEU), handling a third of the UK's container traffic with North America. It was also the second busiest UK port for agricultural products (18 per cent of UK traffic). Ro-ro traffic to Ireland is important, almost 90 per cent of which was with Belfast or Dublin in 2004. Ferry passengers to and from Ireland and the Isle of Man totalled almost 740,000 and a further 600,000 passengers travelled on the Mersey ferry.

TABLE 3.12 Freight at Lancashire and Cumbrian ports, 1965 to 2004

Million tonnes

	1965	1970	1975	1980	1985	1990	1995	2000	2001	2002	2003	2004
Liverpool	31.7	28.8	23.4	12.3	10.4	23.2	30.0	30.4	30.3	30.4	31.7	32.2
Manchester	15.8	16.0	14.5	12.7	9.5	8.1	8.4	7.7	7.9	6.3	6.1	6.6
Heysham	3.6	0.6	0.6	0.9	0.9	1.5	2.7	3.7	3.8	3.7	4.1	3.5
Fleetwood	0.3	0.3	1.0	1.9	1.9	1.4	1.2	1.5	1.6	1.5	1.6	1.7
Garston	1.5	1.9	1.1	1.1	1.5	0.8	0.8	0.5	0.5	0.4	0.4	0.5
Workington	0.9	1.0	0.2	0.3	0.6	0.5	0.6	0.6	0.4	0.4	0.3	0.2
Barrow	0.3	0.1	0.1	–	0.1	0.2	0.3	0.2	0.2	0.3	0.2	0.2
Silloth	0.1	0.1	0.1	0.1	0.1	0.1	0.1	0.2	0.1	0.1	0.2	0.2
Lancaster	..	0.1	–	0.1	0.2	0.2	0.1	0.1	0.1	0.1	0.2	0.1
Preston	..	2.2	1.2	0.7	–	–	–	–	–	–	–	–
Whitehaven	0.4	0.6	0.5	0.7	0.5	0.6	–	–	–	–	–	–
Other ports	..	..	..	0.3	–	–	–	–	–	–	–	–
Total[1]	55.0	52.0	43.0	31.3	25.6	36.5	44.2	45.0	45.0	43.3	44.7	45.2

1 Estimated for earlier years where figures for some ports not available

Source: DfT

TABLE 3.13 Port traffic at Liverpool, 2004

Million tonnes

	Imports	Exports	Total	Domestic	Total
Liquid bulk	8.2	0.1	8.3	3.9	12.2
of which: crude oil	7.7	–	7.7	3.7	11.4
Dry bulk	5.9	2.4	8.3	0.3	8.7
Containers	2.0	2.1	4.2	0.3	4.5
Mobile self-propelled vehicles	0.4	0.6	1.1	0.7	1.8
Mobile non-self-propelled vehicles	1.0	1.5	2.5	1.6	4.2
Other cargo	0.9	–	0.9	–	0.9
Total	18.5	6.8	25.3	6.9	32.2

Source: DfT

Manchester

The area covered by the port is adjacent to the Port of Liverpool and includes the Manchester Ship Canal and the Rivers Weaver, Irwell and upper parts of the Mersey. The port of Manchester handled 6.6 million tonnes in 2004, including oil products (53 per cent) and miscellaneous liquid bulk (24 per cent). Between 1970 and 1990, traffic fell by a half to 8.1 million tonnes, in part due to improved transport links between the region and the rest of the UK. The Ship Canal, which opened in 1894, is approximately 58 kilometres in length and rises over 50 feet from its beginning near Eastham to Salford.

Heysham

Traffic at Heysham totalled 3.5 million tonnes in 2004, of which 80 per cent was in unaccompanied trailers. After falling sharply from 3.6 million to 0.6 million tonnes between 1965 and 1970, traffic at the port has since recovered to its present level. The main destinations are to Ireland and the Isle of Man. There were also 287,000 ferry passengers in 2004.

Fleetwood

Fleetwood handled 1.7 million tonnes of ro-ro traffic and 73,000 ferry passengers in 2004. Traffic is mainly to Larne but also to the Isle of Man. Fish landings were valued at £0.9 million in 2004.

Other ports

Other ports in the region contributed 1.2 million tonnes in 2004. These included Garston (0.5 million), Workington (0.2 million), Barrow (0.2 million), Silloth (0.2 million) and Lancaster (0.1 million). Traffic has been in decline at a number of these ports – for example, in 1970, Preston handled 2.2 million, Garston 1.9 million and Workington 1.0 million tonnes. There is no longer any commercial traffic at Preston or Whitehaven, although fishing at the latter port was valued at £1.8 million in 2004.

Scotland (west coast)

Traffic at Scotland's west coast ports totalled 22 million tonnes in 2004. The dominant ports are Clydeport (52 per cent) and Glensanda (24 per cent), whilst ro-ro traffic is also important at Loch Ryan. Overall, the region has experienced steady growth in traffic since the low of 11 million tonnes in the early 1980s (Table 3.14).

Clydeport

The port authority of Clydeport includes the ports of Ardrossan, Clydebank, Finnart, Glasgow, Greenock and Hunterston. The port has experienced steady growth in recent years, although it has changed considerably from when it was one of the UK's busiest ports and its shipyards built a significant proportion of the world's ships.

TABLE 3.14 Freight at Scotland (west coast) ports, 1965 to 2004												
	Million tonnes											
	1965	1970	1975	1980	1985	1990	1995	2000	2001	2002	2003	2004
Clydeport	15.3	18.0	13.6	8.8	9.9	8.9	7.6	7.2	11.1	9.7	9.2	11.5
Glensanda	–	–	–	–	–	3.3	4.9	5.9	5.5	5.8	5.3	5.2
Cairnryan	..	..	..	0.9	1.1	1.1	2.0	2.3	2.0	2.1	2.3	2.8
Stranraer	–	0.4	1.0	1.0	1.5	1.7	1.9	1.5	1.4	1.3	1.3	1.3
Ayr	1.0	0.6	0.6	0.8	0.8	1.2	0.6	0.3	0.3	0.2	0.3	0.4
Other ports	..	..	..	0.1	0.2	0.3	0.3	0.4	0.8	0.9	0.9	0.8
Total[1]	17.0	19.0	16.0	11.8	13.4	16.6	17.2	17.6	21.0	20.1	19.3	22.0

1 Estimated for earlier years where figures for some ports not available

Source: DfT

Clydeport handled 12 million tonnes in 2004, including coal (53 per cent) and crude oil and oil products (30 per cent). With regard to coal, it handled 15 per cent of the UK total. Container and cruise traffic are also important, whilst various local passenger services operate to nearby islands. The recent upturn in fortunes compares with the sharp fall from 18 million tonnes in 1970 to 5 million tonnes in the early 1990s.

Glensanda

Glensanda's port facility exists primarily for the shipping of stone from the adjacent quarry. The port handled 5.2 million tonnes in 2004, which was almost entirely outward. From its beginning in the late 1980s, traffic at Glensanda had reached almost 6 million tonnes by 2000.

Cairnryan

Cairnryan handled 2.8 million tonnes of ro-ro traffic in 2004, two and a half times the level in 1990. The sailing to Larne is the only significant port traffic, which included 595,000 ferry passengers in 2004. Cairnryan competes with its neighbour Stranraer for traffic and passengers to the north of Ireland.

Stranraer

Stranraer handled 1.3 million tonnes of ro-ro traffic in 2004. The port experienced strong growth between 1970 and 1995, when it reached 1.9 million tonnes, before falling back to the present level. Traffic in 2004 included 1.3 million ferry passengers.

Other ports

Other ports on the west coast of Scotland include Ayr (0.4 million tonnes in 2004), Troon (0.3 million tonnes and 423,000 ferry passengers) and Stornoway (0.2 million tonnes). Traffic at Ayr has declined since handling 1.2 million tonnes in 1990. Fishing is important at Ullapool (£10.5 million of fish landings in 2004), Lochinver (£7.9 million) and Kinlochbervie (£7.6 million) amongst others.

Scotland (east coast)

Ports around Scotland's east and north-east coasts handled 88 million tonnes in 2004, the second highest regional tonnage in the UK. Traffic was dominated by Forth (39 per cent of total traffic), Sullom Voe (27 per cent) and Orkney (20 per cent). Around 80 per cent of all traffic was crude oil. The region experienced huge growth during the 1970s and 1980s – from 8.5 million tonnes in 1965 to 115 million tonnes by 1985 (Table 3.15). This part of Scotland also has a number of the UK's leading fishing ports.

Forth

Forth covers the Firth of Forth area including Braefoot Bay, Grangemouth, Hound Point, Leith and Rosyth. In 2004, the port handled 35 million tonnes, of which 88 per cent was liquid bulk. This total included 13 per cent of the UK's crude oil and almost half of its

TABLE 3.15 Freight at Scotland (east coast) ports, 1965 to 2004

						Million tonnes						
	1965	1970	1975	1980	1985	1990	1995	2000	2001	2002	2003	2004
Forth	6.1	8.3	8.4	28.8	29.1	25.4	47.1	41.1	41.6	42.2	38.8	34.9
Sullom Voe	–	–	–	28.5	59.0	36.0	38.3	38.2	31.2	29.4	26.4	23.9
Orkney	..	..	0.4	15.4	16.1	8.6	12.9	22.8	18.4	18.8	14.4	17.9
Aberdeen	1.3	1.3	1.5	2.0	2.5	3.9	3.6	3.4	3.8	3.6	3.2	3.9
Cromarty Firth	0.2	0.3	1.3	0.8	3.1	1.4	2.3	2.3	2.1	2.7	3.5	3.2
Dundee	0.5	1.0	0.9	1.0	0.9	1.3	1.1	1.0	1.1	1.1	1.0	1.1
Montrose	..	..	0.4	0.5	0.7	0.6	0.7	0.7	0.7	0.7	0.8	0.8
Inverness	0.2	0.3	0.4	0.5	0.8	0.6	0.7	0.7	0.7	0.7	0.7	0.7
Peterhead	–	–	0.6	0.8	1.4	1.6	1.3	1.1	1.3	1.3	1.1	0.7
Lerwick	..	..	0.7	0.9	1.0	0.7	0.9	0.5	1.0	0.7	0.6	0.6
Perth	0.1	0.2	0.2	0.1	0.3	0.3	0.2	0.3	0.2	0.2	0.1	0.2
Other ports	..	..	0.3	0.4	0.4	0.4	0.6	0.6	0.6	0.7	0.6	0.6
Total[1]	8.5	11.5	15.2	79.7	115.3	81.0	109.7	112.9	102.8	102.1	91.2	88.4

1 Estimated for earlier years where figures for some ports not available

Source: DfT

TABLE 3.16 Port traffic at Forth, 2004

	Million tonnes				
	Imports	Exports	Total	Domestic	Total
Liquid bulk	0.8	20.2	21.0	9.8	30.8
of which: crude oil	–	15.5	15.5	5.4	20.9
Dry bulk	0.8	0.1	0.9	0.1	1.0
Containers	0.7	1.1	1.8	0.1	1.9
Mobile self-propelled vehicles	0.1	0.1	0.3	–	0.3
Mobile non-self-propelled vehicles	0.1	0.1	0.2	–	0.2
Other cargo	0.6	0.1	0.7	–	0.8
Total	3.1	21.8	24.9	10.0	34.9

Source: DfT

liquid gas. There are also important container and ro-ro services which carried 215,000 TEU and 192,000 sea passengers in 2004. (Table 3.16)

Two step changes have occurred in Forth's port traffic levels, the first during the mid-1970s when North Sea oil began to be piped through to Houndpoint terminal and a second in the mid-1990s as more fields became connected by pipeline.

Sullom Voe

Sullom Voe in the Shetland Isles handled 24 million tonnes in 2004. This was almost entirely crude oil, for which it was the UK's second busiest port (15 per cent of the total). After beginning in 1978, traffic doubled to 60 million tonnes between 1980 and 1984 with the North Sea oil boom. By 1990, however, traffic had fallen to 36 million tonnes.

Orkney

Orkney, which includes the ports of Flotta, Kirkwall, Scapa Flow, Stromness and Westray, handled 18 million tonnes of traffic in 2004. It is almost totally reliant on crude oil, with remaining traffic making up only 1 per cent of the total. As recently as 2000, Orkney handled 23 million tonnes. Together with the Shetland Isles, sea passengers to and from the mainland totalled 339,000 in 2004.

Aberdeen

Aberdeen is a key base, providing support to the nearby offshore oil industry. The port handled 3.9 million tonnes in 2004, including oil products (35 per cent) and general cargoes (33 per cent). Traffic rose from 1.5 million tonnes in 1975 to almost 4 million tonnes in 1990. Aberdeen is also important for fishing, with £14 million of fish landed in 2004.

Cromarty Firth

Cromarty Firth handled 3.2 million tonnes in 2004, with crude oil making up more than 90 per cent of traffic. Traffic has more than doubled since 1990. The Firth, a natural deep-water harbour, is used by cruise liners and as a base for oil rig maintenance and repair.

Other ports

Other ports in the region include Dundee (1.1 million tonnes), Montrose (0.8 million), Inverness (0.7 million), Peterhead (0.7 million) and Lerwick (0.6 million). Peterhead (£64 million of fish landed in 2004), Fraserburgh (£35 million), Lerwick (£34 million) and Scrabster (£16 million) are leading UK fishing ports.

Domestic ferry services

Domestic ferry services in Scotland provide vital economic and social links. In the west of Scotland, 6.6 million passengers travelled on services provided by Caledonian MacBrayne and Western Ferries in 2004. A further 1.4 million used the Orkney, Shetland and Northlink services (Table 3.17 and Figure 3.5).

TABLE 3.17 Scottish domestic ferry routes handling around 100,000 or more passengers in 2004	
Operator and route	**Passengers ('000)**
Caledonian MacBrayne	5,311
of which:	
Wemyss Bay–Rothesay	764
Largs–Cumbrae Slip	683
Ardrossan–Brodick	717
Oban–Craignure	653
Gourock–Dunoon	620
Colintraive–Rhubodach	268
Fionnphort–Iona	257
Ullapool–Stornoway	189
Mallaig–Armadale	188
Kennacraig–Islay	148
Lochaline–Fishnish	123
Orkney Ferries	322
of which: Kirkwall–Stronsay	98
Shetland Island Council	755
of which:	
Toft–Ulsta, Yell	232
Lerwick–Bressay	213
Laxo, Vidlin–Symbister, Whalsay	154
Gutcher, Yell–Belmont, Unst	123
Western Ferries (Gourock–Dunoon)	1,255
NorthLink Ferries	289
of which: Scrabster–Stromness	143
Total	7,932
Source: Scottish Executive	

FIGURE 3.5 Domestic ferry services in Scotland

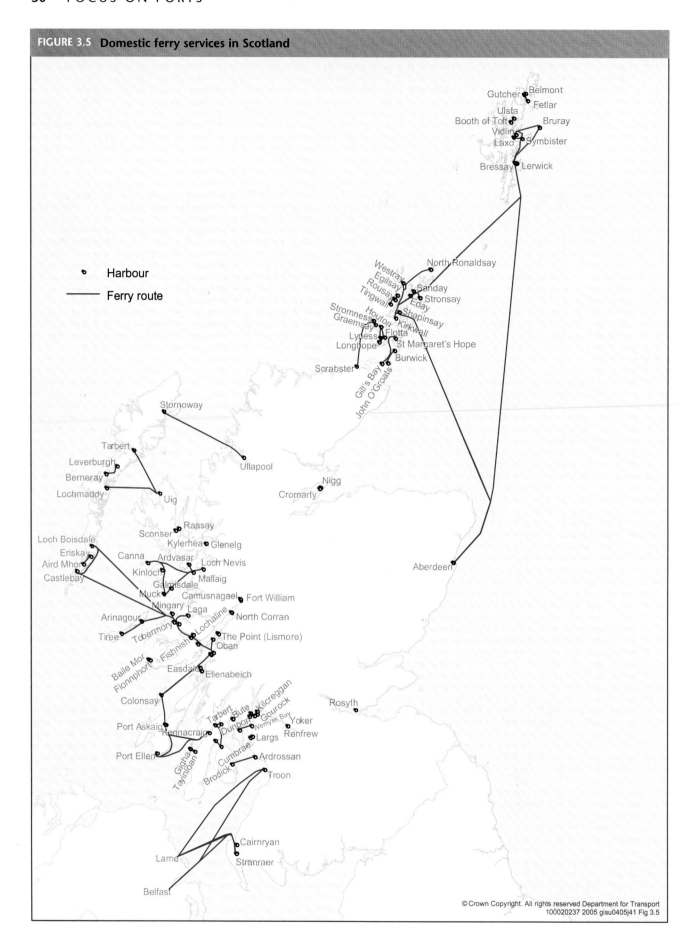

Harbour

Ferry route

North-east

The north-east region handled 59 million tonnes in 2004, of which over 90 per cent was at Tees & Hartlepool and almost two-thirds was liquid bulk. The sharpest growth in the region occurred between 1975 and 1980, when traffic rose by 63 per cent to 52 million tonnes (Table 3.18). Whilst shipbuilding was once a major activity in the region, nowadays ship repair and projects for the offshore industry and Navy predominate.

TABLE 3.18 Freight at North-East ports, 1965 to 2004

						Million tonnes							
	1965	1970	1975	1980	1985	1990	1995	2000	2001	2002	2003	2004
Tees & Hartlepool	12.1	22.7	20.2	39.4	30.6	40.2	46.1	51.5	50.9	50.4	53.8	53.8
Tyne	9.0	7.6	5.6	5.7	5.4	6.3	4.1	2.4	2.5	2.7	2.8	3.0
Sunderland	3.3	2.3	1.9	2.8	2.5	2.5	1.2	0.9	1.0	0.9	1.0	1.1
Blyth	5.4	3.5	3.4	3.5	3.8	1.7	1.2	0.9	0.8	0.8	0.9	0.9
Seaham	1.3	0.5	0.7	0.5	0.9	0.8	0.6	0.5	0.5	0.3	0.5	0.4
Berwick	..	0.1	0.2	0.1	0.1	0.1	0.2	0.1	0.1	0.1	0.1	0.1
Whitby & Scarborough	0.1	0.1	–	0.2	0.2	0.2	0.1	–	–	–	–	–
Warkworth	0.2	–	–	–	–	–	–	–	–	–	–	–
Total[1]	31.5	37.0	32.0	52.1	43.5	51.8	53.4	56.4	55.7	55.2	59.1	59.3

1 Estimated for earlier years where figures for some ports not available

Source: DfT

Tees & Hartlepool

Tees & Hartlepool includes Billingham, Hartlepool, Middlesbrough, Redcar and Teesport. In 2004, it was the UK's second busiest port, handling 54 million tonnes (including 16 per cent of the UK's crude oil and almost a third of its ores traffic). Tees & Hartlepool experienced particularly strong growth during the late 1970s and to a lesser extent during the 1990s (Table 3.19). Of the 25 million tonnes of crude oil handled in 2004, around 95 per cent was outward traffic.

TABLE 3.19 Port traffic at Tees & Hartlepool, 2004

	Million tonnes				
	Imports	Exports	Total	Domestic	Total
Liquid bulk	4.2	17.2	21.5	14.8	36.2
of which: crude oil	0.3	13.2	13.5	11.8	25.3
Dry bulk	10.5	1.2	11.7	0.8	12.5
Containers	0.6	0.6	1.2	–	1.2
Mobile self-propelled vehicles	0.2	0.1	0.2	–	0.2
Mobile non-self-propelled vehicles	1.4	0.8	2.2	–	2.2
Other cargo	0.4	1.1	1.5	–	1.5
Total	17.2	21.0	38.2	15.6	53.8

Source: DfT

Tyne

Tyne ports covers the area along the River Tyne and includes the ports of Newcastle and North and South Shields. It handled 3.0 million tonnes of cargo in 2004. Movements of ores and miscellaneous dry bulks made up 40 per cent of the total, with ro-ro traffic (including trade vehicles) a further 25 per cent. Overall, freight traffic levels have more than halved since 1990, mostly in coal and petroleum products. In 2004, 767,000 sea passengers travelled to and from Scandinavia and northern Europe, whilst fish landings at North Shields totalled £2.8 million.

Sunderland

Sunderland handled 1.1 million tonnes of traffic in 2004, including liquid bulk (60 per cent) and general cargo (30 per cent). This figure compares with total port traffic of 2.5 million tonnes in 1990, the difference largely due to material dumped at sea.

Blyth

Traffic at Blyth totalled of 0.9 million tonnes in 2004, including imports of raw materials for aluminium production. Traffic generally has been in decline for a number of years, largely due to a fall-off in coal exports (in the early 1960s exports reached almost 7 million tonnes).

Other ports

Other ports in the region include Seaham (0.4 million tonnes in 2004) and Berwick (0.1 million). Whitby and Scarborough had fish landings valued at £2.7 million and £1.7 million respectively in 2004.

Humber

Traffic on the Humber totalled 84 million tonnes in 2004 – the third busiest region in the UK. The leading ports were Grimsby & Immingham (69 per cent of total traffic),

TABLE 3.20	Freight at Humber ports, 1965 to 2004											
						Million tonnes						
	1965	1970	1975	1980	1985	1990	1995	2000	2001	2002	2003	2004
Grimsby & Immingham	8.3	23.7	22.0	22.2	29.1	39.4	46.8	52.5	54.8	55.7	55.9	57.6
Hull	9.4	7.2	4.5	3.8	4.5	6.8	10.0	10.7	10.6	10.3	10.5	12.4
Rivers Hull & Humber	..	..	..	4.1	6.3	7.6	6.4	9.0	7.8	8.9	10.0	9.2
River Trent	..	..	..	2.3	3.8	3.2	3.0	2.4	2.4	2.3	2.3	2.3
Goole	2.2	2.2	1.8	1.4	1.4	1.7	2.3	2.7	2.6	2.3	1.9	2.2
River Ouse	..	..	..	0.5	0.9	1.0	0.6	0.3	0.2	0.2	0.2	0.2
Total[1]	..	39.0	34.5	34.3	46.0	59.7	69.0	77.7	78.5	79.7	80.9	84.0

1 Estimated for earlier years where figures for some ports not available

Source: DfT

Hull (15 per cent) and wharves on the Rivers Hull and Humber (11 per cent). Traffic in the region more than doubled between 1980 and 2000 to 78 million tonnes (Table 3.20).

Grimsby & Immingham

Grimsby & Immingham, which includes the ports of Grimsby, Immingham and Killing-holme, was the UK's busiest port in 2004, with total traffic of 58 million tonnes. Total traffic included oil products (24 per cent) and coal (19 per cent) in 2004. Movements of crude oil and unaccompanied trailers are also important. The port, which experienced steady growth between the late 1970s and 2000, handled a third of the UK's ores traffic and a quarter of coal traffic in 2004 (Table 3.21).

A new deep-water cargo facility, the Humber International Terminal, opened in June 2000. The multipurpose terminal is able to accommodate vessels carrying up to 130,000 tonnes. Fish landings at Grimsby were valued at £3.3 million in 2004 (this compares with £20 million to £30 million in the 1970s and 1980s, when Grimsby was the UK's leading fishing port).

TABLE 3.21 Port traffic at Grimsby & Immingham, 2004					
	Million tonnes				
	Imports	Exports	Total foreign traffic	Domestic	Total
Liquid bulk	7.6	6.9	14.5	9.7	24.3
of which: crude oil	3.2	–	3.2	5.9	9.2
oil products	3.5	6.6	10.1	3.6	13.7
Dry bulk	18.1	0.9	19.0	0.2	19.2
of which: Coal	10.4	0.6	11.0	0.1	11.0
Containers	0.7	0.3	1.0	–	1.0
Mobile self-propelled vehicles	1.1	0.7	1.7	–	1.7
Mobile non-self-propelled vehicles	5.7	4.0	9.6	–	9.6
Other cargo	1.5	0.3	1.8	–	1.8
Total	**34.7**	**13.0**	**47.7**	**10.0**	**57.6**
Source: DfT					

Hull

Hull handled 12 million tonnes in 2004, including ro-ro traffic (24 per cent), containers (19 per cent) and miscellaneous liquid bulk (17 per cent). Hull was the UK's sixth busiest container port in terms of TEU in 2004 (310,000). Traffic fell sharply between 1965 and 1980, from 9.0 million to 3.8 million tonnes but had recovered to 10 million tonnes by 1995. Almost 1.0 million sea passengers also passed through Hull in 2004, mainly to Rotterdam and Zeebrugge. In the 1970s, the port was one of the UK's leading fishing ports, with landings valued at £20 million. This had fallen to £7.2 million in 2004, but Hull remains the fourth largest fishing port in England.

Rivers Hull and Humber

The River Hull and River Humber area includes wharves on the north side of the Humber, at Tetney on the south bank of the Humber and in the New Holland area. Total traffic of 9.2 million tonnes in 2004 was dominated by crude oil discharged at the Tetney oil terminal, accounting for 92 per cent of the total.

Other ports

Other regional ports handled 4.7 million tonnes in 2004, including River Trent (2.3 million), Goole (2.2 million) and River Ouse (0.2 million). The various types of cargo handled at these ports include miscellaneous dry bulk, forestry and iron and steel products, general cargo and containers. The River Trent port area includes the wharves at Flixborough and Gunness, and the River Ouse has wharves at Howdendyke and Selby. Traffic at the latter has declined from 1.0 million tonnes in 1990.

Wash and northern East Anglia

Total traffic in the Wash and northern East Anglia of 2.9 million tonnes in 2004 was the smallest regional total, representing around 1 per cent of UK traffic. King's Lynn and Boston each handled a quarter, whilst Great Yarmouth and Sutton Bridge handled a fifth of total tonnage. Traffic in 2004 was about three-quarters of the previous year's tonnage. Traffic in the region peaked around 1990 at just under 7 million tonnes (Table 3.22).

TABLE 3.22 Freight at Wash and northern East Anglian ports, 1965 to 2004												
					Million tonnes							
	1965	1970	1975	1980	1985	1990	1995	2000	2001	2002	2003	2004
King's Lynn	0.8	0.7	1.0	1.0	1.0	1.2	1.0	1.1	0.9	1.0	1.1	0.7
Boston	0.6	0.5	0.7	1.0	1.4	1.3	1.1	1.3	0.8	0.8	1.0	0.7
Great Yarmouth	1.1	1.3	1.4	1.4	2.4	3.0	1.8	0.8	0.7	0.7	0.8	0.6
Sutton Bridge	–	–	–	–	–	0.6	0.9	0.8	0.7	0.7	0.7	0.6
Lowestoft	0.2	0.2	0.3	0.4	0.5	0.5	0.5	0.4	0.3	0.3	0.4	0.2
Wisbech	0.1	0.1	0.4	0.2	0.2	0.1	0.1	0.1	0.1	0.1	–	0.1
Other ports	..	..	..	0.1	0.2	0.1	–	0.1	0.1	–	–	–
Total[1]	2.9	2.9	3.9	4.0	5.7	6.8	5.3	4.5	3.5	3.5	4.0	2.9

1 Estimated for earlier years where figures for some ports not available

Source: DfT

King's Lynn

King's Lynn handled 0.7 million tonnes in 2004. Prior to this, traffic had been fairly stable for a number of years. The port generally caters for dry bulk cargoes as well as forestry and iron and steel products.

Boston

Boston, situated on the River Witham, also handled 0.7 million tonnes in 2004, although traffic has tended to fluctuate from year to year. The main cargoes handled are agricultural products, forestry products and iron and steel products. Between 1970 and 1985, traffic grew from 0.5 million to almost 1.5 million tonnes.

Great Yarmouth

Great Yarmouth's port traffic has declined in recent years to 0.6 million tonnes in 2004. However, between 1975 and 1990, traffic more than doubled to 3.0 million tonnes, including 1.3 million tonnes of offshore and 1.1 million tonnes of unitised traffic. Great Yarmouth is a supply centre for the North Sea oil and gas industry.

Other ports

Other ports in the region include Sutton Bridge, situated on the River Nene (0.6 million tonnes in 2004), Lowestoft (0.2 million) and Wisbech (0.1 million). Lowestoft also landed fish worth £0.8 million.

Haven

Total traffic at Haven ports was 31 million tonnes in 2004. Traffic is dominated by Felixstowe (75 per cent of the total), followed by Harwich (14 per cent) and Ipswich (11 per cent). Traffic in the region almost tripled between 1980 and 2000, reaching 37 million tonnes. Container traffic makes up around two-thirds of the total and ro-ro traffic a further quarter (Table 3.23).

TABLE 3.23 Freight at Haven ports, 1965 to 2004												
						Million tonnes						
	1965	1970	1975	1980	1985	1990	1995	2000	2001	2002	2003	2004
Felixstowe	0.7	2.2	4.1	6.5	10.1	16.4	24.1	29.7	28.4	25.1	22.3	23.4
Harwich	0.6	2.1	2.3	3.2	3.9	4.2	3.6	4.0	2.6	3.5	4.3	4.3
Ipswich	1.9	2.5	2.5	2.8	3.8	4.7	3.5	2.9	2.9	3.3	3.9	3.6
Mistley Quay	..	..	0.1	0.2	0.2	0.3	0.2	0.2	0.2	0.1	0.1	0.1
Other ports	..	..	..	0.1	–	–	–	–	–	–	–	–
Total[1]	3.2	6.9	9.0	12.9	18.0	25.7	31.4	36.8	34.1	32.1	30.6	31.4

1 Estimated for earlier years where figures for some ports not available

Source: DfT

Felixstowe

Total traffic at Felixstowe was 23 million tonnes in 2004, of which 84 per cent was containerised. The port is the largest container port in the UK and fifth largest in Northern Europe. In 2004, it handled 2.7 million TEU, representing 34 per cent of the UK total. Around 40 per cent of container traffic was with the Far East and a quarter was with Europe in 2004. The port experienced strong container growth between 1980 and 2000, reaching 1.9 million units (2.8 million TEU) (Figure 3.6). Felixstowe also handled 2.9 million tonnes of ro-ro traffic in 2004.

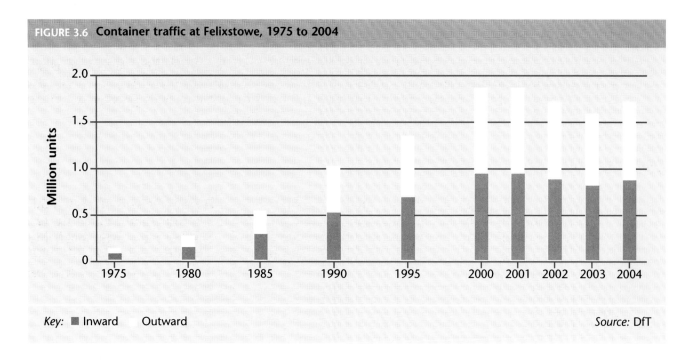

FIGURE 3.6 **Container traffic at Felixstowe, 1975 to 2004**

Key: ■ Inward   Outward

Source: DfT

Harwich

Total traffic at Harwich was 4.3 million tonnes in 2004, of which 85 per cent was ro-ro cargo. After falling to 2.6 million tonnes in 2001, the port now handles a similar amount of tonnage to that handled in 1990. There were also 1.1 million ferry passengers (largely to and from the Netherlands) and 91,000 cruise passengers in 2004.

Ipswich

Ipswich is situated on the River Orwell about 15 kilometres upstream from Felixstowe. It handled 3.6 million tonnes in 2004, including ro-ro cargo (26 per cent) and miscellaneous dry bulk (23 per cent). Agricultural products are also important. Traffic at Ipswich peaked at 5.0 million tonnes in the late 1980s but had fallen to 2.9 million tonnes by 2000.

Mistley Quay, the only other port in the region with commercial traffic, handled 0.1 million tonnes in 2004.

TABLE 3.24 **Freight at Northern Ireland ports, 1980 to 2004**									
				Million tonnes					
	1980	1985	1990	1995	2000	2001	2002	2003	2004
Belfast	5.8	6.0	8.0	10.5	12.5	13.4	12.8	13.2	13.6
Larne	2.8	3.6	4.0	4.7	4.5	3.5	4.3	4.3	5.0
Warrenpoint	0.7	0.8	1.4	1.7	1.7	1.5	1.8	1.9	2.0
Londonderry	0.9	0.8	0.7	1.0	1.1	1.1	1.1	1.2	1.4
Coleraine	0.2	0.2	–	–	–	–	0.1	0.1	0.1
Carrickfergus	0.2	0.3	0.3	–	–	–	–	–	–
Other ports	1.4	1.8	2.1	2.4	1.6	1.7	1.3	1.3	1.4
Total	**12.1**	**13.5**	**16.7**	**20.3**	**21.4**	**21.2**	**21.4**	**22.0**	**23.4**

Source: DfT

Northern Ireland

Ports in Northern Ireland handled 23 million tonnes in 2004. The leading ports are Belfast (58 per cent of total traffic in 2004), Larne (21 per cent), Warrenpoint (8 per cent) and Londonderry (6 per cent). Between 1980 and 2000, traffic grew by more than half to 21 million tonnes, mainly at Belfast (Table 3.24).

Belfast

The port of Belfast handles around a quarter of all Ireland's seaborne trade. In 2004, this amounted to 14 million tonnes, including ro-ro traffic (34 per cent) and oil products (21 per cent) (Figure 3.7). Between 1985 and 2000, traffic at Belfast more than doubled to 13 million tonnes. Around two-thirds is domestic traffic to and from the rest of the UK. There were also 1.8 million ferry passengers in 2004 and the port has seen growth in the number of cruise calls.

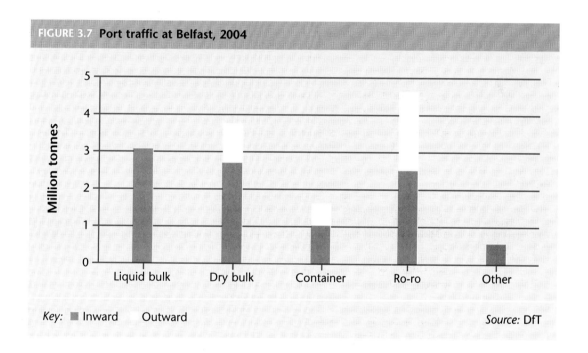

FIGURE 3.7 Port traffic at Belfast, 2004

Key: ■ Inward   Outward

Source: DfT

Larne

Total traffic at Larne was 5.0 million tonnes in 2004, almost entirely ro-ro traffic with Great Britain. Traffic increased by almost 70 per cent between 1980 and 1995 to 4.7 million tonnes, but had fallen to 3.5 million tonnes by 2001. There were also 786,000 ferry passengers at Larne in 2004.

Warrenpoint

Warrenpoint handled 2.0 million tonnes of traffic in 2004. More than 60 per cent was ro-ro traffic with Heysham and 14 per cent imported agricultural products. Traffic at the port more than doubled between 1985 and 1995.

Londonderry

Londonderry handled 1.4 million tonnes in 2004, including coal (39 per cent), oil products (19 per cent) and agricultural products (16 per cent). Total traffic, which is almost entirely inward, increased by almost 20 per cent during 2004.

Other ports

Other ports in Northern Ireland handled 1.5 million tonnes in 2004, including Ballylumford (1.1 million), Kilroot (0.3 million) and Coleraine (0.1 million). Fishing is important at Kilkeel (£5.6 million landed in 2004), Portavogie (£4.7 million) and Ardglass (£2.9 million).

Port employment and accident rates

Estimates of port-related employment in the UK were published by the Department for Transport in November 2005, based on a survey of port authorities and port-related businesses.[1] Previous estimates of port employment were made in 1989 and 1992 by the British Port Federation, and in earlier years dating back to the mid-1960s by the National Ports Council. Although broad trends in employment are evident, the latest results are more widely based and not directly comparable to earlier work.

Over the past few decades, port employment has contracted significantly even though traffic has been increasing, a consequence of increased mechanisation and containerisation. Forty years ago freight traffic through UK ports was estimated to be around 330 million tonnes, increasing by three-quarters to 573 million tonnes in 2004. At the start of this period, there were around 140,000 port employees, of which 65,000 were registered dock workers engaged in loading and unloading cargoes. Port industry employment levels now are only about 40 per cent of the level 40 years ago, and the number of dock workers involved in cargo operations are only a third of the level 40 years ago.

Port communities are made up of a wide range of different businesses and activities ranging from those directly involved in port operations and supporting functions, for example in port management and administration, dredging, vessel mooring, cargo operations, port regulatory services, engineering and maintenance, forwarding agents, shipping agents, importers and exporters and so on. There are other businesses that supply goods and services to those directly related to port operations. There are businesses which are located on ports because it suits their business activity, and other businesses which simply rent or lease premises or land and otherwise have nothing to do with the port. Employment also results from spending by people employed in ports. Separate employment estimates are available for the above categories, which are defined as follows:

- *Direct employment* – employment associated with the main operation of a port and supporting activities
- *Indirect employment* – employment associated with the provision of goods and services to direct employers
- *Partially related employment* – employment by businesses located at a port because it is convenient for their operation, for example they may import raw materials or export finished goods, and wish to have a manufacturing base close to the reception of raw materials or export of goods
- *Unrelated employment* – those businesses on the port estate that are completely unrelated to the activity of the port – they may simply rent or lease port land
- *Induced employment* – employment supported by spending by households of direct and indirect employees.

---

1 Department for Transport, November 2005: *Port Employment and Accident Rates, Transport Statistics Bulletin* SB(05)32

TABLE 4.1 UK port employment, 2004		
	Total employees[1]	Range
Employment on the port estate		
Direct	54,000	44,500–63,500
Indirect	1,000	500–2,000
Partially related	7,000	4,500–9,500
Unrelated	9,500	8,000–11,000
Employment off the port estate		
Direct	19,500	13,000–26,500
Indirect		14,000–43,000
Induced employment		13,000–47,000
1 Full-time equivalents		
Source: DfT		

Table 4.1 gives the employment estimates, in terms of full-time equivalents (ftes). Ranges are also given (95 per cent statistical confidence intervals, or ranges derived from other studies). It is estimated that 54,000 employees (45,000–64,000) work in directly related jobs in ports and around 20,000 employees (13,000–27,000) work in directly related jobs outside the port. In total, 74,000 (58,000–90,000) employees work directly on port-related activities in and outside the port.

There are others working in ports, supplying goods and services (1,000 employees, 500–2,000); or in partially related jobs, for example manufacturing, where the companies use the port, for example to import or export goods (7,000 employees, 5,000–10,000); or in jobs which have nothing to do with the port, for example leasing premises or land (10,000 employees, 8,000–11,000).

Indirect employment in companies supplying goods and services to direct companies is estimated to be in the range 14,000–43,000 people, and induced employment (employment associated with expenditure of those who derive incomes from direct and indirect companies) is estimated to be in the range 13,000–47,000 people.

Non-permanent employees and seasonal variation

Around 4 per cent of port business employees are non-permanent (at quiet times when the survey was conducted), rising to 9 per cent at busy times of the year. These are averages for all job functions; for certain jobs the proportion of non-permanent employment is greater, for example 10 per cent of stevedores are non-permanent at quiet times and this proportion is likely to be much larger at busy times.

Port employment by occupation

Table 4.2 gives the proportions employed by occupation, for direct employment in and outside the port. Around half of direct employment in the port is work in process, plant and machinery operations or in skilled trades, and just less than a third in management and administration. As far as employment outside the port is concerned, about a third of

TABLE 4.2 UK port employment[1] by occupation, 2004		
	% employees on port	% employees off port
Managerial and professional	16	23
Administration, secretarial	14	26
Personal service, sales and customer service	7	16
Skilled trade	19	18
Process, plant and machinery operations	28	12
Other port employment	16	5
Total	100	100
1 Direct employment		
Source: DfT		

employees are in process, plant and machinery operations or in skilled trades, about half in management and administration, and 16 per cent in customer services. The higher management and administration and customer services reflects the type of companies in this group (principally freight forwarders, shipping agents and so on).

Port employment by function

Table 4.3 gives information about employment by type of function, divided into three main operational areas: marine, cargo, passenger, and other. Marine operations include activities such as harbour masters, pilots, dredging, vessel mooring; cargo operations include cargo handlers and warehousing; passenger operations include information officers, traffic marshals security staff; and other operations include port management and administration, port regulatory services engineering and maintenance.

Total employment in all these operational activities totalled 42,000, around 80 per cent of direct employment in ports. The difference between operational and other types of jobs is accounted for by jobs such as forwarding agents, shipping agents, importers and exporters. More than half of operational employment is involved in cargo operations, around a fifth in marine operations, and 5 per cent in passenger operations. The remainder is accounted for by management and administration, engineering and maintenance and regulation.

TABLE 4.3 UK port employment[1] by function, 2004	
	% employees
Marine operations	21
Cargo operations	53
Passenger operations	5
Other operations	21
All port operations	100
1 Direct employment associated with the main operations of a port	
Source: DfT	

Accident rates

Estimates have been made of accident rates in ports separately for:

- direct employees
- non-direct employees
- all employees.

Table 4.4 summarises the accident rates for these groups. The ranges given in brackets are derived using the employment range estimates. The accident rate for direct employment is estimated to be 1.2 per 100 employees on average, annually (1.2 per cent). If the upper and lower ranges for direct employees are used in the calculation, then the accident range is 1.0–1.5 per cent. The accident rate for non-direct employment on port is 0.5 per cent (range 0.4–0.6 per cent).

The accident rate for direct businesses on port calculated from the survey is lower than estimated by Port Skills and Safety Ltd (PSSL), the port industry's organisation for health, safety skills and standards. PSSL estimates the accident rate, based on returns from members, to be 2.8 per cent in 2004. The reason for the difference is likely to be that the PSSL surveys a narrower range of port employment activities and includes companies more directly involved with port operations (cargo handlers for example), where employees are more at risk.

TABLE 4.4 UK port accidents and accident rates, 2004					
	Number of accidents and rate				
	Fatal	**Major**	**>3 day**[1]	**Total**	**Rate per 100 employees**
Accidents to employees of direct companies on port	2	140	510	652	1.2 (1.0–1.5)
Accidents to employees of other[2] companies on port	2	21	58	81	0.5 (0.4–0.6)
Accidents to employees of all companies on port	4	161	568	733	1.0 (0.9–1.3)

1 An over 3 day injury which is not major but which results in an absence from work of more than 3 days; 2 Indirect, partially related or unrelated employment

Source: DfT

Appendix A

This appendix lists ports and harbours in the UK by status: company and private ports, trust ports, or municipal and other publicly operated ports. The list does not claim to be a comprehensive record of every single port in the UK. All commercially significant ports are included, as well as a number of smaller ports and harbours around the UK coast. However, very small wharfs, terminals and harbours may not be included, and if they are included then the list may only include the location of a wharf or terminal rather than the name of the company or operational name. In some cases, for clarity, major port groupings are also included as well as the individual ports which make up the group. The ports and harbours are also shown on the maps in succeeding appendices (but note that where a port is in private as well as public or trust ownership the port is only shown on the maps as public or trust).

Company and private ports in England

Alnmouth	Howdendyke
Axmouth	Hull
Barking	Hull & Humber Rivers
Barrow-in-Furness	Hythe Hard
Beadnell	Hythe Pier
Bembridge	Immingham
Beaulieu River	Ipswich
Billingham	Isle of Grain
Birkenhead	Keadby
Blakeney	Killingholme
Boscastle	Kingsnorth
Boston	Liverpool
Boulmer	Lowestoft
Brancaster Staithe	Lymington Pier
Brighton	Lympstone
Bristol (Avonmouth & Portbury)	Lytham Pier
Bromborough	Manchester Ship Canal
Burnham Overy Staithe	Medway
Burton-upon-Stather	Middlesbrough
Canvey Island	Mistley
Charlestown	Morston Creek
Chatham	Mullion

Christchurch
Cliffe
Clovelly
Combe Martin
Coryton
Craster
Cullercoats
Dagenham Dock
Dartford
Deptford
Dean Point Quarry
Duddon River
Dutch River Wharf
Eastbourne
Eastham
Ellesmere Port
Erith
Exmouth
Falmouth Docks
Fareham
Fawley
Felixstowe
Fleetwood
Flixborough
Folkestone
Fosdyke
Garston
Glasson Dock
Goole
Gorran Haven
Gravesend
Greenhithe
Greenwich
Grimsby
Grove Wharves
Gunness
Hartlepool
Harwich International
Harwich Docks
Hayle
Heysham

Neap House
New Holland
Newhaven
Northfleet
Par
Penberth
Pentewan
Plymouth (Millbay Docks)
Porthleven
Porthoustock
Portland
Portreath
Purfleet
Redcar
Royal Docks Management Authority
Runcorn
Ryde, Isle of Wight
Seaforth
Seaham
Selby
Sheerness
Silloth
Silvertown
Southampton
St Mawes Pier & Harbour
St Michael's Mount
Stanlow
Sutton Bridge
Sutton Harbour, Plymouth
Swanage
Teesport
Teignmouth
Tetney Terminal
Thamesport
Tilbury
Tranmere
Trent River
Trevethoe
Weston Point
Widnes

Company and private ports in Wales

Aberdaron	Pembrey
Barry	Penarth
Cardiff	Port Dinorwic
Colwyn Bay	Port Penrhyn
Fishguard	Port Talbot
Holyhead	Porthclais
Llandudno	Porthdinllaen
Llandulas	Porthgain
Mostyn	Solva
Newport	Swansea

Company and private ports in Scotland

Ardrishaig	Grangemouth
Ardrossan	Granton
Ayr	Greenock
Bowling	Hound Point
Braefoot Terminal	Hunterston
Burntisland	Inverkeithing
Cairnryan	Kirkcaldy
Clydebank	Leith
Clydeport	Lossiemouth
Corpach	Methil
Dundee	Newburgh
Finnart	Port Glasgow
Forth	Rosyth
Gill's Bay	Stranraer
Glasgow	Tayport
Glensanda	Troon

Company and private ports in Northern Ireland

Cloghan
Kilroot
Larne
Magheramorne

Trust ports in England

Berwick-upon-Tweed
Blyth
Bridlington
Brightlingsea
Cattewater Harbour Commissioners, Plymouth
Chichester
Cowes, Isle of Wight
Crouch Harbour Authority
Dart Harbour & Navigation Authority
Dover Harbour Board
Falmouth Harbour Commissioners
Flamborough
Fowey
Gloucester
Great Yarmouth
Harwich Haven Authority
Hope Cove Harbour Commissioners
King's Lynn Conservancy Board
Lancaster
Langstone
Littlehampton
Looe
Lymington
Maldon
Maryport Harbour Commissioners
Mevagissey
Mousehole
Newlyn
North Sunderland Harbour Commissoners
Orford
Padstow
Polperro Harbour Trustees
Poole
Port Isaac
Port of London
Portloe
Sandwich
Sennan Cove
Shoreham
Staithes
Swanage Pier Trust
Teignmouth Harbour Commissioners
Tyne
Warkworth Harbour Commissioners
Wells
Whitehaven
Yarmouth (IoW) Harbour Commission
Yealm

Trust ports in Wales

Caernarfon
Milford Haven
Neath Harbour Commission
Newport
Pembroke (Milford Docks)
Saundersfoot

Trust ports in Scotland

Aberdeen
Cromarty Firth
Eyemouth
Fraserburgh
Invergordon
Inverness
Lerwick
Mallaig
Montrose
Peterhead
Scrabster
St Margaret's Hope
Stornoway
Tarbert
Ullapool
Wick

Trust ports in Northern Ireland

Belfast
Coleraine
Donaghadee

Londonderry
Warrenpoint

Municipal and other publicly operated ports in England

Appledore
Barnstaple
Bideford
Bridgwater
Bridport
Bristol City Docks
Brixham
Broadstairs
Bude
Coverack
Exeter
Hamble
Harrington
Ilfracombe
Knightstone (Weston-super-Mare)
Leeds (British Waterways)
Leigh-on-Sea
Lydney (Environment Agency)
Lyme Regis
Margate
Minehead
Morecombe
Newbiggin
Newport, Isle of Wight
Newquay
Norfolk Broads
Norwich
Paignton
Penryn
Penzance

Plymouth
Portscatho
Portsmouth
Portwrinkle
Preston
Ramsgate
River Dee (Environment Agency)
River Ouse (British Waterways)
Ryde, Isle of Wight
Rye Harbour (Environment Agency)
Salcombe
Sandown Pier
Scarborough
Seaton Sluice
Sharpness, Glos. (British Waterways)
Southend
Southwold
St Ives
Sunderland
Topsham
Torquay
Truro
Uphill
Watchet
Weymouth
Whitby
Whitstable
Wisbech
Workington
Worthing

Municipal and other publicly operated ports in Wales

Aberaeron
Aberdovey (Aberdyfi)
Aberystwyth

Haverfordwest
Kidwelly
Llanelli

Amlwch

Barmouth

Beaumaris

Burry Port

Cardigan

Carmarthen

Chepstow

Conwy

Menai Bridge

New Quay

Porthcawl

Portmadog

Pwllheli

Rhyl

Tenby

Municipal and other publicly operated ports in Scotland

Arbroath

Armadale

Baltasound

Barra Castlebay

Buckie

Burghead

Burray Pier

Campbeltown

Coll

Colonsay

Craignure

Cumbrae

Dunoon

Eday

Egilsay

Eigg

Faslane

Flotta Terminal

Girvan

Gourock

Graemsay

Kennacraig

Kilchoan

Kinlochbervie

Kirkcudbright

Kirkwall

Kyle of Lochalsh

Largs

Laxo

Lismore

Lochaline

Lochinver

Lochmaddy

Longhope

Lyness

Macduff

North Ronaldsay

Oban

Out Skerries

Papa Westray

Perth

Pittenweem

Port Askaig

Port Ellen

Portree

Renfrew

Rothesay

Sanday

Scalloway

Scapa Flow

Shapinsay

Stonehaven

Stromness

Stronsay

Sullom Voe

Symbister

Thurso

Tingwall

Tiree

Tobermory

Uig

Wemyss Bay

Wyre

Municipal and other publicly operated ports in Northern Ireland

Annalong

Ardglass

Ballintoy

Ballycastle

Ballyhalbert

Ballylumford

Ballywalter

Bangor

Carnlough

Carrickfergus

Church Bay, Rathlin

Cushendall

Dundrum

Glenarm

Groomsport

Kilkeel

Killough

Killyleagh

Kircubbin

Newcastle

Newry

Portaferry

Portavogie

Portballintrae

Portrush

Portstewart

Red Bay

Strangford

Appendix

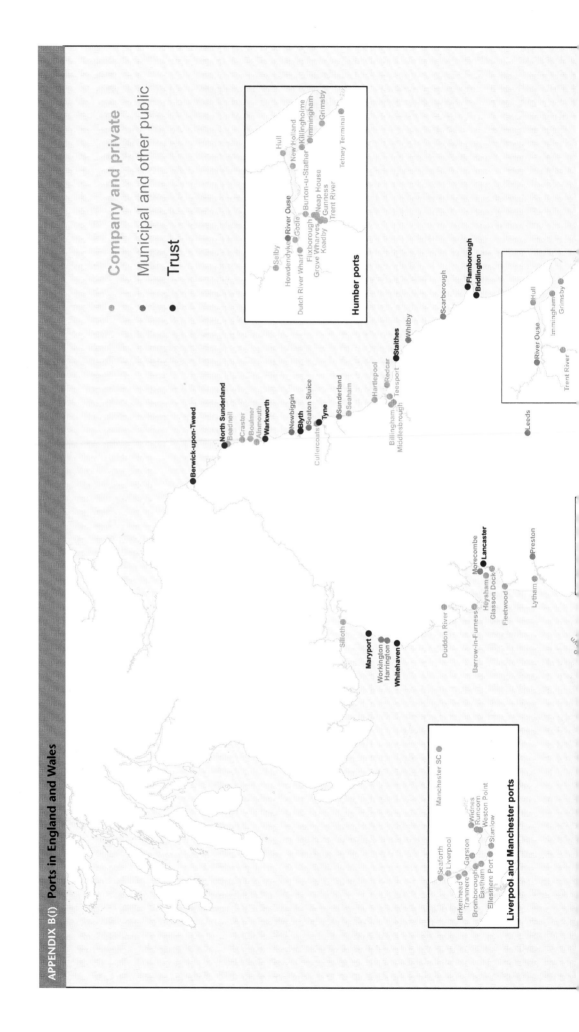

Company and private

Municipal and other public

Trust

Humber ports

Liverpool and Manchester ports

71

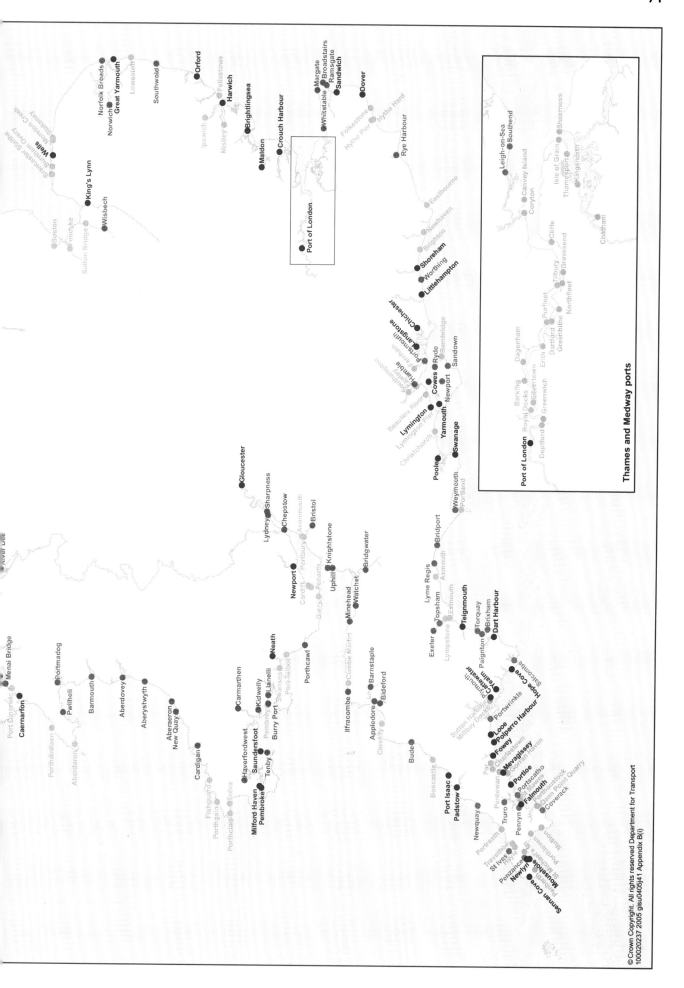

Thames and Medway ports

72

Company and private

Municipal and other public

Trust

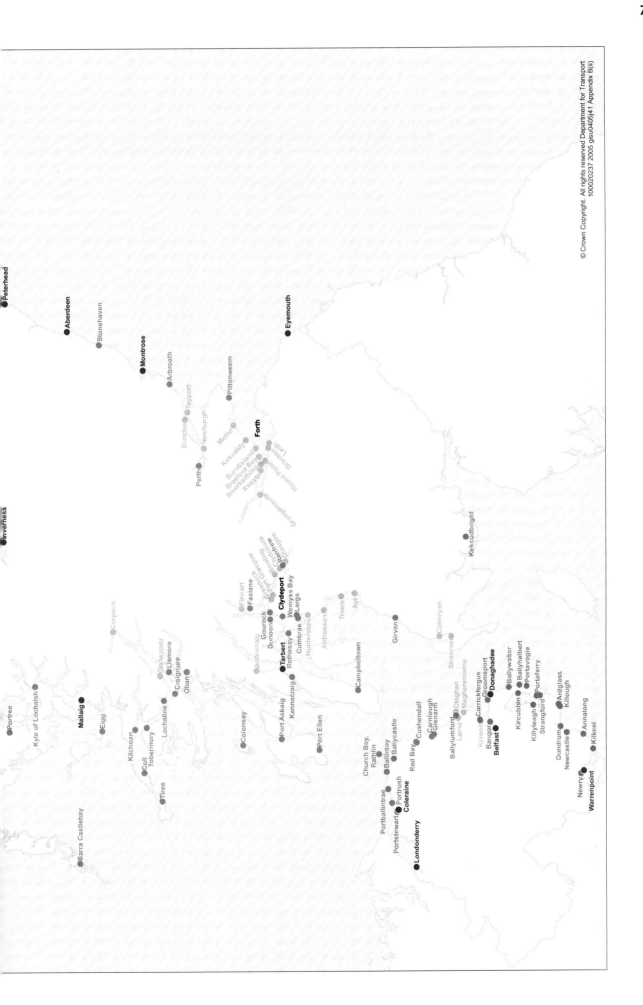

Appendix

APPENDIX C(i) Company and private ports in England and Wales

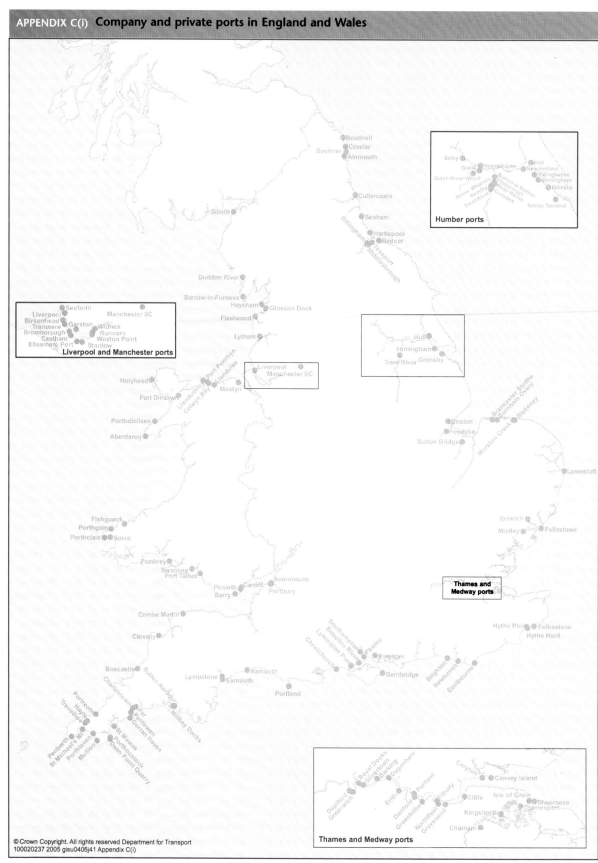

APPENDIX C(ii)   Company and private ports in Scotland and Northern Ireland

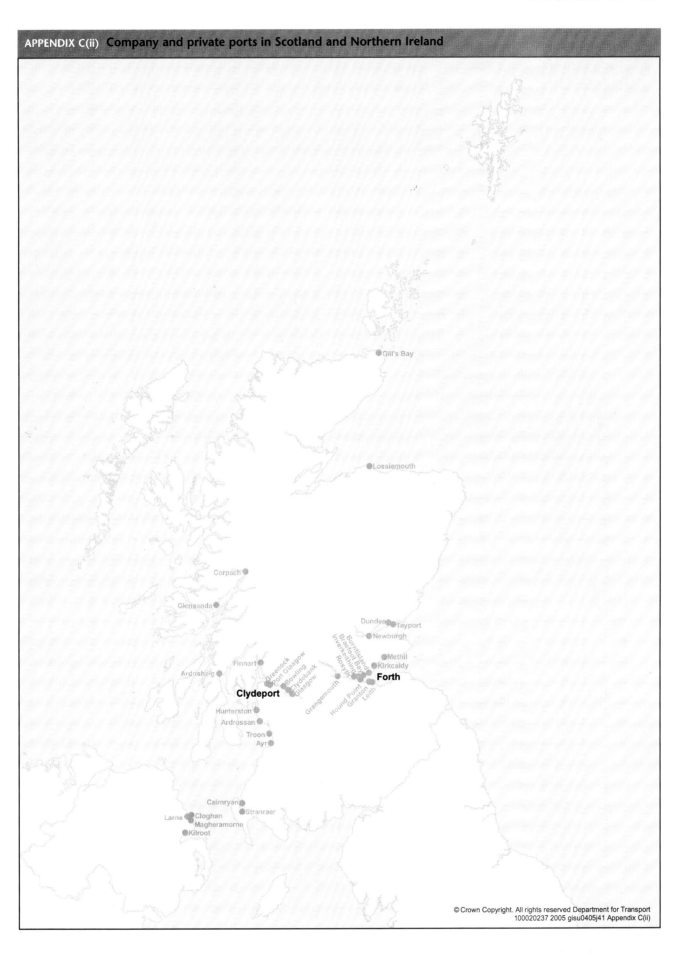

Gill's Bay

Lossiemouth

Corpach

Glensanda

Dundee Tayport
Newburgh

Burntisland
Braefoot Bay
Inverkeithing
Rosyth

Finnart
Greenock
Port Glasgow
Bowling
Clydebank
Glasgow

Ardrishaig

Methil
Kirkcaldy

Forth

Grangemouth
Hound Point
Granton
Leith

Clydeport

Hunterston

Ardrossan

Troon
Ayr

Cairnryan
Stranraer
Larne Cloghan
Magheramorne
Kilroot

Appendix

Appendix E

APPENDIX E(i) Municipal and other publicly owned ports in England and Wales

APPENDIX E(ii)   Municipal and other publicly owned ports in Scotland and Northern Ireland

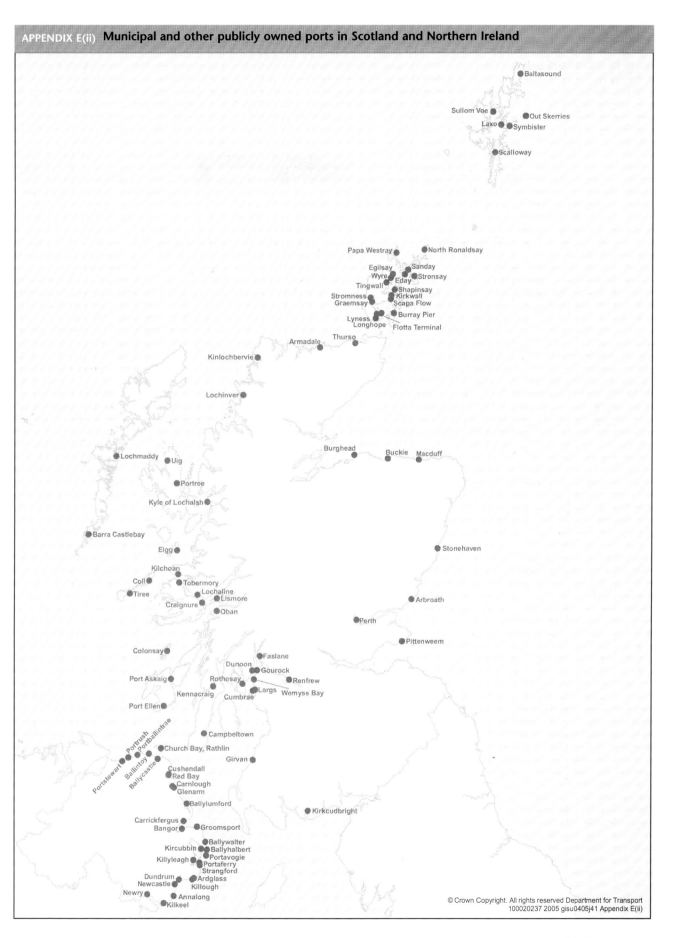

Appendix

This appendix lists all commercially active ports in the UK which supply traffic returns to the Department, and also the name of the authority or undertaking. These returns provide the basis for most of the information which is presented in this publication. For statistical purposes ports are grouped into 'major' ports, which may comprise several ports within a group. The authorities which make returns for these major ports are marked with an asterisk in the list below. Major ports are, in most cases, those ports which handle at least one million tonnes of freight traffic annually.

List of port authorities and undertakings supplying port traffic returns for 2004

Port group	Port (* major port)	Authority/undertaking
Thames and Kent	Brightlingsea	Brightlingsea Port Ltd
	Wallasea	Baltic Distribution
	London*	Port of London Authority
	Medway*	
	(inc. Thamesport)	Medway Ports Ltd
	Whitstable	Canterbury City Council
	Ramsgate*	Thanet District Council
	Dover*	Dover Harbour Board
	Folkestone	Folkestone Properties Ltd
	Other ports	Port of Rye
Sussex and Hampshire	Newhaven*	Newhaven Port and Properties Ltd
	Shoreham*	Shoreham Port Authority
	Littlehampton	Littlehampton Harbour Board
	Portsmouth*	Portsmouth Commercial Port
	Southampton*	Associated British Ports
	Southampton*	Southampton Container Terminals Ltd
	Cowes IOW	Cowes Harbour Commissioners
	Other ports	Newport Harbour (Isle of Wight)
West Country	Poole*	Poole Harbour Commissioners
	Teignmouth	Associated British Ports
	Plymouth*	Associated British Ports
	Plymouth*	Cattewater Harbour Commissioners
	Fowey*	Fowey Harbour Commissioners
	Par	South Coast UK Ltd
	Falmouth	Falmouth Harbour Commissioners
	Falmouth	A & P Falmouth Ltd
	Falmouth	Falmouth Oil Services (1994) Ltd

Port group	Port (* major port)	Authority/undertaking
	Other ports	Weymouth and Portland Borough Council
	Other ports	Portland Port Ltd
	Other ports	Torbay Borough Authority
	Other ports	Dart Harbour and Navigation Authority
	Other ports	Carrick District Council – Truro
	Other ports	Carrick District Council – Penryn
	Other ports	RMC Aggregates Ltd – Dean Point
	Other ports	Penzance Harbour Authority
	Other ports	Newlyn Pier and Harbour Commissioners
	Other ports	Padstow Harbour Commissioners
	Other ports	West of England Quarry, Porthoustock
Bristol	Bridgwater	Sedgemoor District Council
Channel	Bristol*	Bristol Port Company
	Gloucester and Sharpness	British Waterways
	Newport*	Newport Harbour Commissioners
	Newport*	Associated British Ports
	Cardiff*	Associated British Ports
	Barry	Associated British Ports
	Port Talbot*	Associated British Ports
	Neath	Neath Harbour Commissioners
	Swansea*	Associated British Ports
	Other ports	O J Williams (Texaco) – Yelland
	Other ports	Torridge District Council – Bideford
West and	Milford Haven*	Milford Haven Port Authority
North Wales	Fishguard*	Stena Line Ltd
	Holyhead*	Stena Line Ports Ltd
	Mostyn	Port of Mostyn
	Other ports	North West Aggregates Ltd – Llanddulas
	Other ports	Port Penrhyn Plant Ltd
Lancashire	Liverpool*	Mersey Docks & Harbour Company
and Cumbria	Liverpool*	Mersey Wharf – Bromborough
	Garston	Associated British Ports
	Manchester*	Manchester Ship Canal Company
	Fleetwood*	Associated British Ports
	Lancaster	Lancaster Port Commissioners
	Heysham*	Heysham Port Ltd
	Barrow	Associated British Ports
	Workington	Cumbria County Council
	Silloth	Associated British Ports
Scotland:	Stranraer*	Stena Line Ltd
west coast	Cairnryan*	P&O European Ferries Ltd
	Ayr	Associated British Ports
	Clyde*	Clydeport Operations Ltd
	Glensanda*	Foster Yeoman Ltd
	Other ports	Associated British Ports – Troon
	Other ports	British Waterways – Ardrishaig & Corpach

Port group	Port (* major port)	Authority/undertaking
	Other ports	Tilcon (Scotland) Ltd – Lochaline
	Other ports	Stornoway Pier & Harbour Commission
Scotland: east coast	Orkney*	Orkney Islands Council
	Lerwick	Lerwick Port Authority
	Sullom Voe*	Shetland Islands Council
	Cromarty Firth*	Cromarty Firth Port Authority
	Inverness	Inverness Harbour Trust
	Peterhead*	Peterhead Bay Authority
	Peterhead*	Peterhead Harbour Trustees
	Aberdeen*	Aberdeen Harbour Board
	Montrose	Montrose Port Authority
	Dundee*	Port of Dundee Ltd
	Perth	Perth & Kinross Council
	Forth*	Forth Ports plc
	Other ports	Wick Harbour Trust
	Other ports	Scrabster Harbour Trust
	Other ports	The Moray District Council – Buckie
	Other ports	Aberdeenshire Council – Macduff
	Other ports	The Moray District Council – Burghead
	Other ports	Fraserburgh Harbour Commissioners
	Other ports	Tilcon (Northern) Ltd – Inverkeithing
	Other ports	Inverkeithing Port Services
North-east	Berwick	Berwick Harbour Commission
	Blyth	Port of Blyth
	Tyne*	Port of Tyne Authority
	Sunderland*	Port of Sunderland Authority
	Seaham	Seaham Harbour Dock Company
	Tees and Hartlepool*	PD Teesport
	Whitby and Scarborough	Scarborough Borough Council – Whitby
Humber	Hull*	Associated British Ports
	Rivers Hull & Humber*	New Holland Dock (Wharfingers) Ltd
	Rivers Hull & Humber*	New Holland Bulk Services Ltd
	Rivers Hull & Humber*	John L Seaton & Co. Ltd – Hull
	Rivers Hull & Humber*	Karlshamns Ltd
	Rivers Hull & Humber*	Conoco UK Ltd – Tetney
	River Trent*	Flixborough Wharf Ltd
	River Trent*	Gunness Wharf Ltd
	River Trent*	Trenship Agency Ltd – Neap House Wharf
	River Trent*	J Wharton (Shipping) Ltd – Grove Wharf
	River Trent*	PD Port Services – Keadby
	Goole*	Associated British Ports
	River Ouse	PD Port Services – Howdendyke
	River Ouse	General Freight Company – Selby
	Grimsby & Immingham*	Associated British Ports
	Grimsby & Immingham*	Associated Petroleum Terminals (Immingham) Ltd
	Grimsby & Immingham*	Humber Sea Terminal

Port group	Port (* major port)	Authority/undertaking
	Grimsby & Immingham*	Simon Cargo Ltd
	Other ports	Storefreight Services Ltd – Dutch River Wharf
Wash and northern East Anglia	Boston*	Port of Boston Ltd
	Wisbech	Fenland District Council
	Sutton Bridge	Port Sutton Bridge Ltd
	King's Lynn	King's Lynn Conservancy Board
	Great Yarmouth*	Great Yarmouth Port Authority
	Lowestoft	Associated British Ports
	Other ports	Wells Harbour Commissioners
Haven	Felixstowe*	Felixstowe Dock and Railway Company Ltd
	Felixstowe*	Maritime Cargo Processing Ltd
	Ipswich*	Associated British Ports
	Mistley Quay	Mistley Quay & Forwarding Co. Ltd
	Harwich*	Harwich International Port Ltd – Parkeston Quay
	Harwich*	Harwich Dock Co. Ltd – Navyard Wharf
Northern Ireland	Londonderry*	Londonderry Port & Harbour Commissioners
	Coleraine	Coleraine Harbour Commisioners
	Larne*	Larne Harbour Ltd
	Carrickfergus	Carrickfergus Borough Council
	Belfast*	Belfast Harbour Commissioners
	Warrenpoint*	Warrenpoint Harbour Authority
	Other ports	Irish Salt Mining & Exploration Co. Ltd – Kilroot
	Other ports	Kilroot Power Ltd